UNDERSTANDING INNOVATION IN EMERGING ECONOMIC SPACES

The Dynamics of Economic Space

Series Editor: Christine Tamásy, University of Vechta, Germany

The IGU Commission on 'The Dynamics of Economic Space' aims to play a leading international role in the development, promulgation and dissemination of new ideas in economic geography. It has as its goal the development of a strong analytical perspective on the processes, problems and policies associated with the dynamics of local and regional economies as they are incorporated into the globalizing world economy. In recognition of the increasing complexity of the world economy, the Commission's interests include: industrial production; business, professional and financial services, and the broader service economy including e-business; corporations, corporate power, enterprise and entrepreneurship; the changing world of work and intensifying economic interconnectedness.

Other books in this series

Economic Development in Rural Areas
Edited by Peter Dannenberg and Elmar Kulke
ISBN 978-1-4724-4481-3

Regional Resilience, Economy and Society
Edited by Christine Tamásy and Javier Revilla Diez
ISBN 978-1-4094-6848-6

Local Food Systems in Old Industrial Regions
Edited by Neil Reid, Jay D. Gatrell, and Paula S. Ross
ISBN 978-1-4094-3221-0

Industrial Transition
Edited by Martina Fromhold-Eisebith and Martina Fuchs
ISBN 978-1-4094-3121-3

Missing Links in Labour Geography
Edited by Ann Cecilie Bergene, Sylvi B. Endresen and Hege Merete Knutsen
ISBN 978-0-7546-7798-7

Globalising Worlds and New Economic Configurations
Edited by Christine Tamásy and Mike Taylor
ISBN 978-0-7546-7377-4

Understanding Innovation in Emerging Economic Spaces

Global and Local Actors, Networks and Embeddedness

Edited by

GRZEGORZ MICEK

Jagiellonian University in Kraków, Poland

Routledge
Taylor & Francis Group
LONDON AND NEW YORK

First published 2015 by Ashgate Publishing

2 Park Square, Milton Park, Abingdon, Oxfordshire OX14 4RN
711 Third Avenue, New York, NY 10017

Routledge is an imprint of the Taylor & Francis Group, an informa business

First issued in paperback 2018

British Library Cataloguing in Publication Data
A catalogue record for this book is available from the British Library.

The Library of Congress has cataloged the printed edition as follows:
Understanding innovation in emerging economic spaces : global and local actors, networks and embeddedness / by Grzegorz Micek.
pages cm. – (The dynamics of economic space)
Includes bibliographical references and index.
ISBN 978-1-4724-1033-7 (hardback) 1. Economic development—Developing countries. 2. Entrepreneurship--Developing countries. 3. Technological innovations--aspects--Developing countries. 4. Information technology--Economic aspects--Developing countries. 5. Research, Industrial--Developing countries. 6. Diffusion of innovations--Developing countries. I. Micek, Grzegorz.

HC59.7.U466 2015
338′.064091724–dc23

2014049937

ISBN 978-1-4724-1033-7 (hbk)
ISBN 978-1-138-54724-7 (pbk)

Contents

List of Figures

List of Tables

Notes on Contributors

Sascha Brinkhoff is a PhD candidate at the Department of Geography at the Humboldt-Universität zu Berlin. In his research he mainly focuses on the impact of knowledge management in territoral innovations systems and creative industries in metropolitan regions. Furthermore, he works as a policy consultant in regional economic development projects in Germany and internationally.

Piotr Dawidko is a PhD candidate at the Institute of Geography and Spatial Management of the Jagiellonian University in Kraków, Poland. His scientific interests include geography of high technology industries, regional innovation systems and the role of global production networks in the restructuring of post-communist economies.

Magdalena Dej graduated from the Jagiellonian University in Kraków, where she received her PhD in the earth sciences/geography. She is a researcher at the Institute of Urban Development in Kraków. She specializes in local and regional development as well as rural geography. Magdalena Dej is the author and co-author of over 30 research papers concerned with the issue of large companies and their impact on local communities, rural areas, local labor markets, and foreign investment. She has participated in several international research projects financed by the European Union and the International Visegrad Fund. Currently, she is in charge of a research project concerning the relocation and expansion of business enterprises in Polish metropolitan areas.

Sławomir Dorocki graduated from the Pedagogical University of Cracow with an MA degree in geography and a PhD in history (Institute of European Studies of the Jagiellonian University in Kraków). He is Associate Professor at Pedagogical University of Cracow, Institute of Geography. His research interests are tied with regional problems and processes of socio-economic regionalization, with particular emphasis on the diversity of Europe, processes of European integration and historical conditions.

Sylvi Birgit Endresen is Associate Professor at the Institute of Sociology and Human Geography, University of Oslo. Her main research area is economic geography, with an emphasis on technological change and labor issues. Her recent publications include contributions in *Missing Links in Labor Geography* (co-edited with A.C. Bergene and H.M. Knutsen, Ashgate 2010); and 'More than just a Production Factor: The View of Labor in the Works of the Norwegian

Economist Torkel Aschehoug (1822–1909)' (co-authored with C.M. Fasting), in *The Liberation of the Serfs: The Economics of Unfree Labor* (edited by J. Backhaus; Springer 2012), as well as the forthcoming article in *International Encyclopedia of Social and Behavioural Sciences*, Elsevier, 'The Geography of Labour'.

Aksel Ersoy is a Postdoctoral research assistant at the University of Bristol. He received his PhD in 2012 in the School of Geography, Earth and Environmental Sciences at the University of Birmingham where he worked as a research assistant in the 7th EC Program entitled, Shrink Smart: The Governance of Shrinkage. Before joining University of Bristol, he worked as a teaching fellow in Spatial Planning at the University of Aberdeen. His research interests lie at the interface between local and regional economic development, urban planning, policy design and the connected communities programme in the UK. He is a member of the Regional Studies Association, American Association of Geographers and Royal Geographical Society. He also holds an Associate Fellowship of Higher Education of Academy.

Maciej Huculak is a social and economic geographer. He graduated from Jagiellonian University and had been working as a staff member at the Institute of Urban Development in Kraków. His research interests include the beer brewing industry in Poland. He is an author of many research papers in the area of local and regional development including papers on special economic zones, labor markets, urban renewal, and recently the development of recreational geothermal bathing facilities. Maciej Huculak has been a member of many research teams working on projects at the national and international level. He was the coordinator of the following project: Local Economic Development in Airport Catchment Areas—AirLed. The project was financed by the Central Europe—European Union Programme.

Wojciech Jarczewski is a social and economic geographer. He is an Associate Director for Research at the Institute of Urban Development in Kraków. In both his theoretical and applied research projects, he is intrigued by the following question: How can local, regional, and national governments revive and support social and economic development? He has sought to answer this question by pursuing national and international research projects that address the following issues: (1) ways for townships to attract investors, (2) geographic factors associated with the establishment of new plants, (3) evaluation of project effectiveness in the case of projects and programs co-financed by the European Union, (4) urban renewal, (5) creation of new townships, (6) development of recreational geothermal bathing facilities. His research has been published in several dozen research papers and more than a dozen books.

Egert Juuse is a PhD student and junior research fellow at the Chair of Innovation Policy and Technology Governance, Tallinn University of Technology, Estonia.

His main research areas are financial policies and innovation policies in catching-up economies, especially Central and Eastern Europe. He has been involved in several national research projects, e.g. Public Administration and Development in Small States; Innovation Policies and Uneven Development. Currently he is involved in the 7th EU Framework Programme Project FESSUD: Financialization, Economy, Society and Sustainable Development (2011–2016).

Rainer Kattel is a Professor and Chair of Innovation Policy and Technology Governance, Tallinn University of Technology, Estonia. His main research area is industrial and innovation policies in catching-up economies, especially Central and Eastern Europe and Latin America. He has published extensively on innovation policy and development economics. His recent books include *Ragnar Nurkse: Trade and Development* (co-edited with Jan A. Kregel and Erik S. Reinert; Anthem, 2009), *Knowledge Governance: Reasserting the Public Interest* (co-edited with Leonardo Burlamaqui and Ana Celia Castro; Anthem, 2012) and *Public Procurement for Innovation Policy: International Perspectives* (co-edited with Tarmo Kalvet and Veiko Lember; Springer, 2014).

Elmar Kulke is a Professor of Economic Geography at the Department of Geography and the Dean of the Natural Sciences Faculty at the Humboldt-Universität zu Berlin. His major research areas are the Geography of Services in a variety of sectors in Central Europe and South-East Asia besides innovation driven case studies in Technology Parks. He has authored and edited many books; among recent publications are *Wirtschaftsgeographie* (2013, 5th Edition) and *Wirtschaftsgeographie Deutschlands* (2010, 2nd Edition).

Michał Męczyński is a lecturer and researcher in the Institute of Socio-Economic Geography and Spatial Management at the Adam Mickiewicz University in Poznan. He received his PhD in earth sciences/geography in 2006. In 2006–2010 he was a postdoctoral researcher on the project: Accommodating Creative Knowledge—Competitiveness of European Metropolitan Regions within the Enlarged Union financed by the 6th EU Framework Programme. His main research interests include: development of the labor market in the creative sector, mobility and entrepreneurship of creative class representativeness, and the role of innovations in the economy.

Grzegorz Micek received his PhD in the earth sciences/geography in 2006. As a member of Regional Studies Association he coordinated the RSA Research Network: Regional Development and the Changing Geography of Outsourcing Business Services in an Enlarged Europe (2008–2011). Grzegorz Micek was also involved in the 6th EU Framework Programme Project: The Moving Frontier: The Changing Geography in Labour Intensive Industries (2005–2007). His research fields include the local impact of global companies in the IT industry, biotechnology, and business services; proximity economics; applications of knowledge flows and spillovers concepts to regional development theories. He has published in several peer reviewed journals, e.g. *Journal of Economic Geography, Urban Studies, European Planning Studies, Geografiska Annaler: Series B* and *Tijdschrift voor Economische en Sociale Geografie.*

Mathew Novak is an Assistant Professor of Geography at Saint Mary's University in Halifax, Canada . His research analyses the built form of cities and how they change over time. Currently his work focuses on the retail sector, and the social and economic dimensions of consumption landscapes.

Tomasz Rachwał (PhD) is an economic geographer. He is a member of the National Council for Entrepreneurship, Head of Department of Entrepreneurship and Spatial Management of Institute of Geography of the Pedagogical University of Cracow, and a Scientific Secretary of the Industrial Geography Committee of the Polish Geographical Society. Tomasz Rachwał is a Deputy Editor in Chief of the following journals: *Studies of the Industrial Geography Commission of the Polish Geographical Society*, *Entrepreneurship—Education* and *Annales Universitatis Paedagogicae Cracoviensis—Studies Geographica.* His research interests focus primarily on the issue of the change of spatial structures of industry, the functioning of various branches of industrial activity, corporate restructuring and the role of entrepreneurship in local and regional development.

Lech Suwala holds a PhD in economic geography, and is a junior lecturer at the Department of Geography at the Humboldt-Universität and at the University of Applied Sciences of Technology and Economics in Berlin. Furthermore, he worked as a professional in a private policy consulting company for several years. His scientific interests comprise spatial aspects of creativity, innovation and entrepreneurship where he conducted manifold studies at the district, federal and European level.

Łukasz Sykała graduated from the Jagiellonian University in Kraków with a degree in geography and public management. He is currently a PhD student at the Institute of Geography and Spatial Management at Jagiellonian University. He is also a staff member at the Institute of Urban Development in Kraków. His research interests revolve around issues in regional policy, especially the effect of the EU's cohesion policy on local and regional development.

Anna Irena Szymańska graduated from Cracow University of Economics with an MA degree in Management and Marketing and PhD degree in economic sciences in the field of management sciences (Chair of Market Analysis and Marketing Research—Cracow University of Economics in Krakow). She is an Associate Professor in the Department of Entrepreneurship and Spatial Management at the Pedagogical University of Cracow. Her research interests are related to the issue of consumer needs, preferences and market behaviour as well as issues in the area of entrepreneurship and innovation with particular emphasis on the SME sector.

Krzysztof Wiedermann is an Associate Professor in the Department of Socio-Economic Geography at the Pedagogical University of Crakow. His research interests focus on the issues of local and regional development, urban and regional economics, urban restructuring and industrial districts with particular emphasis on the relocation of industry and services, foreign direct investment and instruments in support of economic development—the special economic zones, industrial parks.

Małgorzata Zdon-Korzeniowska graduated from Cracow University of Economics (Department of Economics, specializing in Entrepreneurship and Innovation) with an MA degree in economy and PhD in economic sciences in the discipline of management studies (Faculty of Management and Social Communication, Jagiellonian University). She is Associate Professor at the Pedagogical University of Cracow, Institute of Geography, Department of Entrepreneurship and Spatial Management. Her research interests center around the issues of tourism management at the local and regional level, with particular emphasis on sustainable development and the construction of local and regional tourist products, as well as the integration and coordination of the actors involved in this process, with particular emphasis on the role of local government.

Acknowledgements

I would like to express my gratitude to Neil Reid, Chair of the Commission on The Dynamics of Economic Spaces of the International Geographical Union for his help and support of this project. Great thanks must be extended to the Head of my Department, Bolek Domański, who supported and encouraged me during the reviewing and publishing process. The book could not have been completed without the kindness and calmness of my wife and children. Ela, Ania, Adam – thank you! The usual disclaimer applies – the views expressed are the sole opinions of the authors.

List of Abbreviations

CAP	Common Agricultural Policy
CEE	Central and Eastern Europe
CIT	Corporate Income Tax
CMA	Cracow Metropolitan Area
DBFs	Dedicated Biotechnology Firms
EU	European Union
FDI	Foreign Direct Investments
GDP	Gross Domestic Product
GPN	Global Production Networks
GVA	Gross Value Added
HCLQ	Horizontal Clustering Location Quotient
HEI	Higher Education Institutions
HR	Human Resources
IPRs	Intellectual Property Rights
KM	Knowledge Management
KNM	Knowledge Network Management
LQ	Location Quotient
NIZ	Niepołomice Investment Zone
nMRAs	Non–Metropolitan Rural Areas
PIT	Personal Income Tax
PMR	Poznań Metropolitan Area
R&D	Research and Development
RIS	Regional Innovation Systems
SMEs	Small and Medium Enterprises
SOEs	State-Owned Enterprises
TIS	Territorial Innovation Systems
TNCs	Transnational Corporations

Introduction

Innovation in Emerging Economic Spaces in the Context of the Local–Global Dichotomy

Grzegorz Micek

Dense networks of local inter-connected firms facilitate the creation of an innovative milieu. Some academics argue that formal and informal linkages resulting in knowledge flows between companies and other stakeholders lead to enhanced local innovation performance. Similar processes may be observed at transnational corporations, which gradually become embedded in relationships with local partners (Domański 2004). This is especially true of manufacturing companies, and in particular the automotive industry (Domański and Gwosdz, 2009, Jürgens and Krzywdzinski 2009). In general, new economic spaces of innovation and creativity emerge and consequently lead to geographical, institutional and social changes.

Innovation is widely understood in the book as every product, process, idea or behaviour that is new to a company. This is in line with the definition of innovation by Kotler and Armstrong (2010) who emphasize that innovation is a complex process that may occur in different forms. Innovation perceived in such a way is massively uneven in terms of spatial representation. A limited number of countries, regions, cities, and localities are powerful gatekeepers and generate the bulk of creative and innovative ideas, while the majority is largely left out. The book looks at neglected, but emerging innovation centres analysed from various spatial and organizational perspectives – ranging from entire countries and regions to individual firms and small neighbourhoods. A variety of economic sectors are examined including biotechnology, agrotourism, and the food retail industry. The book employs various, often contradictory, concepts, ranging from local buzz and the global pipeline, through an analysis of collective learning processes to geographical embeddedness. Both qualitative and quantitative approaches are used.

The presented research originates mainly from emerging economies, mostly from Central and Eastern Europe (CEE): Poland and Estonia. CEE has been undergoing economic, social, and political transformation for 25 years. The transition from state-controlled to free market economies has resulted in FDI inflow (Weresa 2004) and rapid growth of local entrepreneurship (Piasecki and Rogut 1993). Some companies have become rising stars of innovation in the region, which was until recently perceived as a peripheral one. In the course of this transition, new industries have emerged; for instance, advanced business services

(Hardy, Micek and Capik 2011, Micek, Działek and Górecki 2011) and the IT industry, in particular (Guzik and Micek 2008). Siłka (2012) provides evidence that the innovation potential of the largest Polish cities has accelerated after EU enlargement.

The aims of the book are twofold. The authors describe changes occurring in the regions and cities under transformation and attempt to find common and unique mechanisms behind these changes. Consequently, scholars shed a light on the scale and scope of the innovativeness of selected economic and social processes.

The book consists of 12 chapters. Part I includes three chapters that examine innovation from slightly different spatial and methodological perspectives. In Chapter 1, Sascha Brinkhoff, Lech Suwala, and Elmar Kulke (Humboldt-Universität zu Berlin) look at innovation on the micro scale (Berlin-Adlershof and Seville-Cartuja science parks) and aim to contribute to the discussion of the interdependence of innovative businesses with scientific institutions in neo-Marshallian industrial districts or 'localities of learning'. The scholars investigate how different dimensions of proximity (including social approaches such as social proximity, embeddedness, and trust) and personal commitment to specific spaces ('ba') contribute to enhanced collective learning processes. Second, Brinkhoff et al. integrate the concept of knowledge management with the scale of territorial innovation networks. Ultimately, the analysis adds new findings to the recipe as to how specific 'knowledge cluster management' instruments and institutions support the building of distinct natures of proximity and further key elements that promote localized learning and innovation.

Chapter 2, by Tomasz Rachwał and Krzysztof Wiedermann (Pedagogical University of Kraków), contains the results of macro-scale quantitive research that focuses on the relationship between the development of industry and the R&D community in the EU. The analysis of EU regional industrial change is based on research on changes in employment, gross value added, productivity, and labour costs, while the consequences of the transformation of industrial structures are measured by the value of investment in fixed assets, R&D growth, and employment rates in the high-tech industry. Rachwał and Wiedermann conclude by stating that many Central and Eastern European regions experience the stage of transition to an innovative economy based upon the development of manufacturing activities, which does not necessarily translate into growth in the R&D sector. Despite the major role that manufacturing plays in innovation and R&D, no clear relationship was observed between industrial spending on innovation, R&D indicators, and economic growth, as measured via GDP. Five distinct paths of R&D growth have been identified in relation to GDP growth in the European Union.

In the last chapter of this section, Grzegorz Micek (Jagiellonian University) explores different terms used to discuss knowledge interactions. By looking at recent academic papers, he states that knowledge spillovers and diffusion are the most appropriate words to describe the majority of cases. It is argued that mixed methods (preferably based on in-depth social network analysis) should be used to find a balance between biased results obtained from empirical modelling and case studies. A critique of a variety of indicators of knowledge interactions is also presented.

Innovation in emerging economies is often implemented by foreign and/or large companies, thus the role of these enterprises is scrutinized in Part II. The significant positive impact of FDI on overall regional innovation capacity or spillover effects (Cheung, Lin 2003, Fu 2008) is widely studied in Asian countries. Weresa (2004) argues that FDI impacts innovation processes in different branches of Poland's industrial sector, except for low-technology industries, and thus helps Poland catch up with the rest of the European Union. However, a considerable body of literature has shown that externally owned corporations fail to generate both local and regional economic success (Braidford et al. 2003, Pavlinek 2004, van Egaraat and Jacobson 2005). The majority of contributions in the second section examine relationships between FDI, large companies, local and regional development, and innovative capacity of regions. This remains in line with studies on the local impact of companies from different industries, not only in the context of growth (Stryjakiewicz 2004, Domański and Gwosdz 2005, Glasson 2005, Rachwał and Wiedermann, 2008, Jarczewski and Huculak 2011, Micek 2011), but also in the context of factory closures (Chapain and Murrie 2008).

Aksel Ersoy (University of Birmingham) investigates, in Chapter 4, how large corporations in Turkey contribute to the formation of national and global economies. He examines the geography of the largest corporations in the Turkish urban system, giving consideration to their location, performance, and concentration. Ersoy argues that not only the spatial organization of enterprises remains important in urban systems, but also how they are linked to the dynamics of local and regional economies in developing countries. He states that local embeddedness of the largest corporations leads to an institutional thickness that is thought to be one crucial success factor for regions in a globalizing economy.

In Chapter 5, Egert Juuse (Tallinn University of Technology), Sylvi Birgit Endresen (University of Oslo), and Rainer Kattel (Tallinn University of Technology) examine local embeddedness and networking at foreign companies operating in food retailing and related industries in Estonia. The analysis is based on interviews with key stakeholders as well on secondary sources of information. The authors argue that FDI in Estonian retail has led to increased concentration of capital, sector imbalances, significant market concentration, and thus created vulnerabilities to systemic risks. With a shrinking population and volatile purchasing power, aggressive expansion of foreign retail chains in Estonia has saturated local markets, in terms of excess supply, and incurred sectoral imbalances as well as systemic risks in an unfettered market economy, that is, unmanaged FDI policy in Estonia's transition to a full-fledged market economy.

In Chapter 6, the same three authors analyse the restructuring of food retailing in Estonia and observe a clear trend of instituting positive feedback mechanisms in the Estonian food industry, which brings forth the dual nature of the economy – niche products from small enterprises marketed either in open markets or small independent eco-stores with a trivial 1–2 per cent market share, and industrial style dominating the retail chains. Juuse, Endresen, and Kattel argue that the emergence of small producers and shops has been enabled by agglomeration economies and facilitated by modern retail formats, which have opened up possibilities for the rejuvenation of small shops selling niche farm products in shopping centres or supermarkets. The development path of the Estonian food retailing industry has been affected by institutional and economic adjustments such as radical change in the nation's government and national ideology in the early 1990s, transformation of the structure of the national economy, as well as related developments such as increased urbanization, acquisition of the country's banking sector by Scandinavian banks, and the completion of Estonia's business privatization process. In addition, the legacy of the Soviet period, small market size and the country's non-corporate, neoliberal, nationalistic socioeconomic environment have had a significant impact on the food retailing industry.

In Chapter 7, empirically sound research by Magdalena Dej (Jagiellonian University) focuses on the impact of large companies on rural communities. The research is based on 12 case studies, statistical analysis of the level of development of all Polish non-metropolitan rural areas, and survey studies conducted on a total of approximately 2,500 inhabitants of rural areas found in the vicinity of large companies. Dej's study confirms the significance of the origin of capital in the degree of impact of large companies on local communities in non-metropolitan rural areas. Companies strongly connected to their geographic location or the place of residence of their owners or the managing personnel satisfy all their personnel needs locally. This includes top-level employees. On the other hand, companies that are external with respect to their geographic location, including companies with foreign capital, rely on higher level employees coming from outside their immediate geographic vicinity. The examined foreign companies, as opposed to Polish companies, are characterized by very weak collaboration with local suppliers and service companies, and tend to limit their local sourcing to transportation companies.

In Chapter 8, Wojciech Jarczewski, Maciej Huculak (Institute of Urban Development), Magdalena Dej, and Łukasz Sykała (Jagiellonian University) conduct research on the economic and spatial advantages and disadvantages of the operation of the industrial park in Niepołomice found in the Kraków Metropolitan Area. The impact of company relocations from Kraków to neighbouring Niepołomice on tax revenues and employment has been studied. Although the research has not denied the economic success observed in Niepołomice, it allowed for an analysis of the scale of its local character. In fact, the success of Niepołomice is a success for the entire Kraków metropolitan region. The findings in the chapter make the success story of Niepołomice slightly more realistic.

The behaviour and strategies of innovative small companies are described in Part III, which starts with a contribution by Matthew Novak (Central Washington University), who examines independent cafés in Washington State in the United States. He looks at the creativity of the coffee shop enterprise itself. The data used to examine the business practices of coffee shops, both corporate and independent, include field observations and a comprehensive GIS database of retail outlets across Washington State. Novak employs field observations to get an indication of the size and scope of the business, as well as the type of customers, and novel advertising and branding techniques employed in attracting patrons to their caffeine-laden offerings. He argues that small coffee shops are prime examples of spatial and functional creativity and innovativeness in contrast to the increasing dominance of several global players. Small independent shops show remarkable resilience in facing their corporate rivals. Their small size, limited capital investment, and flexible organizational structure allow them to succeed in a competitive market.

In Chapter 10, Michał Męczyński (Adam Mickiewicz University) attempts to determine the role played by creativity and innovativeness of creative sector managers in the socioeconomic development of the Poznań metropolitan region. He identifies the problems that restrict the creativity of managers and their companies (e.g. brain drain) and links them with various policies pursued by the city's government. Męczyński argues that Poznań should adopt many solutions already introduced in other cities. An example of this would be the implementation of a policy that would regulate rents on premises that belong to the city and are not leased by creative sector entities. A larger concentration of artists resulting from higher building density in the city centre would be conducive to the flow of ideas and the implementation of collaborative projects.

Piotr Dawidko (Jagiellonian University) analyses in Chapter 11 the global–local dichotomy in the Polish biotechnology industry based on a series of semi-structured interviews. He contrasts the Polish biotechnology industry with its Western European counterparts. Certain aspects of the biotechnology industry (e.g. high embeddedness in local knowledge networks) and the present economic situation (generally low innovativeness of Poland's economy) hamper foreign investment in Poland. Dawidko argues that the concepts of local buzz and global pipeline appear to apply only to a limited number of expansion-oriented, highly innovative companies in Poland. At the same time, their usefulness in explaining innovation and the development of small biotechnology companies is limited. Global knowledge flows and labour mobility affect the specialized aspects of business activity, but local factors such as lack of entrepreneurial culture or financial support negatively impact business performance and act as a constraint to business development.

Finally, in Chapter 12, Sławomir Dorocki, Anna Irena Szymańska, and Małgorzata Zdon-Korzeniowska (Pedagogical University of Kraków) examine entrepreneurial behaviour at family businesses in Poland's agricultural tourism sector. The research is based on in-depth interviews with innovative business owners. The authors claim that family history shapes the evolution of the entrepreneurial spirit. Farms are potential areas of development for family entrepreneurship and innovativeness.

References

Braidford, P., MacKinnon, D., Phelps, N.A. and Stone, I. 2003. Embedding the multinationals? Institutions and the development of overseas manufacturing affiliates in Wales and Northeast England. *Regional Studies*, 37(1), 27–40.

Chapain, C. and Murie, A. 2008. The impact of factory closure on local communities and economies: The case of the MG Rover Longbridge closure in Birmingham. *Policy Studies*, 29(3), 305–17.

Cheung, K. and Lin, P. 2004. Spillover effects of FDI on innovation in China: Evidence from the provincial data. *China Economic Review*, 15, 25–44.

Domański, B. 2004. Local and regional embeddedness of foreign industrial investors in Poland. *Prace Geograficzne Jagiellonian University*, 114, 37–54.

Domański, B. and Gwosdz, K. (eds). 2005. *Dziesięć lat doświadczeń pierwszej polskiej specjalnej strefy ekonomicznej. Mielec 1995–2005*. Kraków-Mielec: Institute of Geography and Spatial Management UJ and ARP.

Domański, B. and Gwosdz, K. 2009. Toward a more embedded production system? Automotive supply networks and localized capabilities in Poland. *Growth and Change*, 40, 452–82.

Egeraat, C. van and Jacobson, D. 2005. Geography of Linkages in the Irish and Scottish Computer Hardware Industry: the Role of Logistics. *Economic Geography*, 81(3), 283–303.

Fu, X. 2008. Foreign Direct Investment, Absorptive Capacity and Regional Innovation Capabilities: Evidence from China. *Oxford Development Studies*, 36(1), 89–110.

Glasson, J. 2005. Better monitoring for better impact management: The local socio-economic impacts of constructing Sizewell B nuclear power station. *Impact Assessment and Project Appraisal*, 23(3), 215–26.

Guzik, R. and Micek, G. 2008. Impact of Delocalisation on the European Software Industry, in *The Moving Frontier: The Changing Geography of Production in Labour Intensive Industries* edited by L. Labrianidis. Farnham: Ashgate, 229–54.

Hardy, J., Micek, G. and Capik, P. 2011. Upgrading Local Economies in Central and Eastern Europe? The Role of Business Service Foreign Direct Investment in the Knowledge Economy. *European Planning Studies*, 9, 1581–91.

Jarczewski, W. and Huculak M. (eds). 2011. *Sukces polityki proinwestycyjnej. Niepołomice 1990–2010. Wnioski dla władz lokalnych.* Kraków: Institute of Urban Development.

Jürgens, U. and Krzywdzinski, M. 2009. Changing east–west division of labour in the European automotive industry. *European Urban and Regional Studies*, 16, 27–42.

Kotler, P. and Armstrong, G. 2010. *Principles of Marketing*, Upper Saddle River, NJ: Pearson Education Inc.

Micek, G., Działek, J. and Górecki, J. 2011. The Discourse and Realities of Offshore Business Services to Kraków. *European Planning Studies*, 9, 1651–68.

Micek, G. 2010. Oddziaływanie firm transportu pasażerskiego na rozwój Krakowskiego Obszaru Metropolitalnego. *Prace Geograficzne Jagiellonian University*, 124, 29–43.

Pavlínek, P. 2004. Regional Development Implications of Foreign Direct Investment in Central Europe. *European Urban and Regional Studies*, 11(1), 47–70.

Piasecki, B. and Rogut, A. 1993. *Self-regulation of SME Sector Development at a more Advanced Stage of Transformation.* Tel Aviv: The 20th Annual Conference of E.A.R.I.E.

Rachwał, T. and Wiedermann, K. 2008. Multiplier effects in regional development: the case of the motor vehicle industry in Silesian voivodeship (Poland). *Quaestiones Geographicae, Series B, Human Geography and Spatial Management*, 27(1), 67–80.

Siłka, P. 2012. Potencjał innowacyjny wybranych miast Polski a ich rozwój gospodarczy, *Geographical Studies, Institute of Geography and Spatial Organisation*, 236.

Stryjakiewicz, T. (ed.). 2004. *Wpływ inwestorów zagranicznych na rozwój regionalny i lokalny na przykładzie GlaxoSmithKline Pharmaceuticals S.A. w Poznaniu*. Poznań: Bogucki Wydawnictwo Naukowe.

Weresa, M.A. 2004. Can foreign direct investment help Poland catch up with the EU? *Communist & Post-Communist Studies*, 37(3), 413–27.

PART I
Introduction

Chapter 1
Managing Innovation in 'Localities of Learning' in Berlin and Seville

Sascha Brinkhoff, Lech Suwala, Elmar Kulke

Introduction

In the knowledge economy, a complex dynamic and interactive learning process among multiple stakeholders characterizes the innovation process. Being aware of this process, regional government policies aim to stimulate and even plan innovation in designated 'localities of learning' (Dosi 1996) and 'sticky places' (Markusen 1996). In this respect, science parks have become a prominent instrument in regional development in the knowledge economy. They play an important role as organizational links ('territorial innovation framework') between the pillars of the 'triple helix', that is public research and higher education institutions (HEI), businesses, and public administration (Etzkowitz and Leydesdorff 2000). However, not only direct linkages between different entities need to be considered inside the territorial innovation frameworks. We argue that topical links between knowledge organizations and space are of high importance, too (Suwala 2013). Consequently, two central questions arise for 'innovation agents' such as science parks and regional policy makers; how social cohesion and a trustful atmosphere is actively created in specific 'localities of learning' and territorial innovation systems, and how knowledge network management systems have to be orchestrated successfully to promote interactive and collective learning, respectively. In this chapter, we apply concepts of knowledge creation and management, currently mainly associated with organizational learning in firms and inter-firm collaboration, to the management of such territorial seedbeds of innovation. Here, we elaborate the impact of distinct dimensions of knowledge network management (KNM) instruments – relational (individual–individual) practices and, introducing a new dimensional strand, primarily topical (individual–space) mechanisms.

The chapter is organized as follows. The next section provides a brief overview of the literature of territorial innovation systems (TIS) and introduces approaches of knowledge management (KM) with an emphasis on the *ba* (*basho*) as a genuine setting for relational and topical learning. Then, we elaborate on our definition and the different conceptual strands of knowledge network management (KNM) on the spatial scale of regional innovation systems (RIS) and 'localities of learning'. The following section presents our case studies and the gathered results. Finally, we draw conclusions.

Theoretical Background

As the theoretical framework, we refer to the well-known concepts of territorial innovation systems (Moulaert and Sekia 2003) and the central findings of the knowledge management and knowledge creation literature and apply them to distinct 'localities of learning' (Dosi 1996). We predominantly consider the spatial scale of specific localities such as science parks as the favourable framework for knowledge creation and learning. The concept of 'localities of learning' explicitly underlines the importance of a localized ecosystem of HEI, research institutions, high-technology firms and skilled talent (Kujath and Stein 2011).[1] These elements also are widely discussed on various geographical scales in related concepts such as innovative milieu (Aydalot 1986), localized production systems (Bouchrara 1987), new industrial spaces (Storper and Scott 1989), clusters of innovation (Porter 1990), regional innovation systems (Lundvall 1988, 1992, Braczyk, Cooke and Heidenreich 1998) and learning regions (Florida 1995, Asheim 1997, Morgan 1997). All of them emphasize the importance of spatial proximity next to locational factors in generating production, knowledge, learning processes and/or innovation (Kulke 2008).

Locational factors like qualified talent, local culture and infrastructure are now considered as embedded into TIS with different stakeholders. Notwithstanding, the firm – and not individuals – still is considered to be the nucleus for the economic progress (Moulaert and Sekia 2003). The 'triple helix' (for example Etzkowitz and Leydesdorff 1997, 2000) is the most popular metaphor for these approaches and takes a closer look at the innovation dynamics based on inter-linkages between the private sector, HEI and R&D centres as well as public administration (Etzkowitz and Leydesdorff 2000, Youtie and Shapira 2008).

However, recent research in territorial innovation systems in general and localities of learning, in particular, focuses on the detailed analysis of the relationship between knowledge, learning, innovation and the interplay of the multi-dimensional set of proximity such as cognitive, social, institutional and organizational proximity (Boschma 2005, Torre and Rallet 2005, Knoben and Oerlemans 2006, Thune 2009, Brinkhoff, Suwala and Kulke 2012). The role of science parks as territorial innovative seedbeds and interfaces between industry, universities and other research institutions is heavily examined (Vedovello 1997, Siegel et al. 2003, Siegel, Waldman and Link 2003, Siegel, Westhead and Wright 2003, Kulke 2008, Brinkhoff, Suwala and Kulke 2012, Huber 2012).

In the global knowledge economy, science parks have to redefine their role within the learning environment substantially. Hansson (2007) identifies several

1 For these reasons and our purpose of investigating science parks in particular, we prefer the notion of 'localities of learning' with its idiosyncratic emphasis on knowledge and learning to other mentioned concepts, which centre production and innovation (for instance localized production systems and innovative milieus) or larger spatial scales (for example learning region).

differences between knowledge organizations and science parks. Knowledge organizations ('child of the knowledge society', Hansson 2007: 362) often focus on intangible knowledge, that is the management of knowledge creation through networking, learning, boundary-crossing, team-based cooperation, as well as internal and external interaction. In contrast, most classical science parks are characterized as 'children of the late industrial society' (Hansson 2007: 362). Here, science parks primarily focus on locational factors, material products and act as intermediary frameworks between companies, products and innovation.

In the knowledge economy, science parks are currently experiencing a revival as organizational links between scientific institutions, the private sector and public administration institutions within the regional 'triple helix' (Hansson 2007). Consequently, modern science parks have to evolve to active coordinating institutions by creating and managing direct relationships to other entities or by promoting indirect linkages among knowledge organizations via space (Suwala 2013). This setting allows the establishment of an enabling context, space and environment, where the creation of *ba* (Nonaka, 1991, 1994, Nonaka and Konno 1998) stimulates inter-organizational links among a variety of knowledge organizations. But what is *ba* exactly?

Essentially, the concept of *ba* describes physical (for example a café), social (for instance conversation), cultural (for example customs), mental (for instance shared personal experiences and ideas), economic (for instance trade) or virtual (for example web chat rooms) spaces or interactive platforms (similar to communities of practice), but entails an idiosyncratic feature that enables knowledge creation. At the same time, *ba* exists at different levels (individuals, groups and organizations). When different *bas* overlap, they amplify and accelerate the process of knowledge creation and constitute a greater *ba*, which is called *basho* (Nonaka and Konno 1998, Senoo et al. 2007, Choo and Alvarenga Neto 2010). Therefore, *ba* represents a shared context, where knowledge is created, shared, and utilized. These settings may evolve into very creative environments and highlight the two characteristics of interaction and space-dependent, enabling contexts (Suwala 2013).

The concepts of *ba* and *basho* have been introduced into management literature by the seminal contributions of Nonaka and colleagues (Nonaka 1991, 1994, Nonaka and Takeuchi 1995, Nonaka and Konno 1998). However, the idea was only – even very extensively – applied to the process of knowledge creation (Haarmakorpi and Melkas 2005, Hansson 2007, Senoo et al. 2007, Hautala 2011). The very fundamentals of *ba* or *basho* as introduced by the Kyoto School of Psychology (Nishida 1970, 1999) and refined from the perspective of systems theory (Shimizu 1995), however, have hardly been investigated. Exceptions only exist in scarcely accessible Japanese literature from Psychology (Latka 2003) and Systems Sciences (Katai et al. 2007). Nevertheless, Nonaka's contributions resulted in a shift of understanding of information processing to a knowledge-creating management within enterprises by overcoming substantial and static conceptions of knowledge. Not surprisingly, the managerial focus of the respective studies put less emphasis on the spatial foundations of the *ba*.

Returning to the *ba*'s two outlined properties of interaction and space-dependency and applying them to the dominating agent-based perspectives in economic geography, we conclude that the *ba* includes the well-examined different natures of proximity (social, cognitive, organizational and geographical) (Boschma 2005, Knoben and Oerlemans 2006). From this perspective, space is constructed as relational. Alternatively, we want to break new ground and introduce a new way of representing space as a topical phenomenon (as an important attribute of *ba*) (Latka 2003, Pfister 2007, Suwala 2013) besides the common absolute, relative, and relational understandings of space (Couclelis 1992). Parts of this topical understanding refer to fundamental behavioural features of Japanese society and its predominant organization into small groups, which constitute 'protected spaces'. There, individuals are indirectly bound to each other via the *ba* (Nishida 1970). This results in prevalent direct relationships of individuals and organizations towards enabling spaces, which we call topical relationships (adapted from Latka 2003).

If we investigate more accurately and take a closer look at the etymology of *ba* and *basho*, it becomes evident why it makes sense to differentiate between relational and topical relationships. *Ba* has a variety of meanings in Japanese. It symbolizes different spaces: a) place, b) built form and seat, c) situation and context, d) scene in a theatre play, e) area and sphere of activity, f) arena, g) market at the stock exchange, and h) physical field (Latka 2003).

The meaning of *basho* also has multiple facets. It is understood as a) place, b) position, c) seat, and d) space. However, the current Japanese understanding of *basho* refers either to a place, where an agency is directed to, or to a location, where something or somebody is located (Latka 2003). While the second meaning stresses a static and substance-based location of existence, the first understanding clearly visualizes an individual-to-space relationship. Figure 1.1 indicates the difference between relational and topical links by means of a poly-centric (relational) and topical-centric (topical) network (Latka 2003, Suwala 2013).

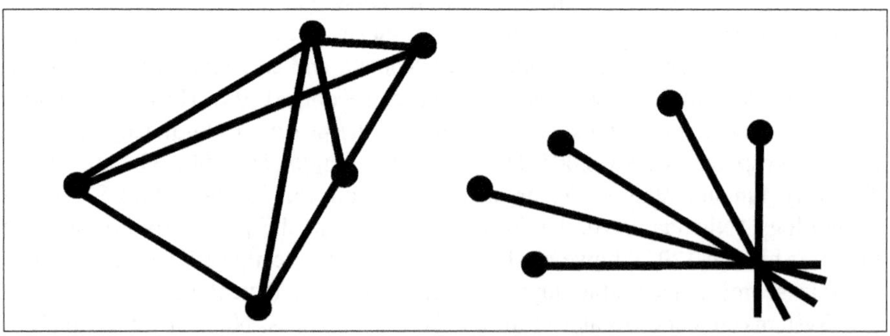

Figure 1.1 Poly-centric (relational, left) and topo-centric (topical, right) networks

Source: own elaboration based on Latka (2003) and Suwala (2013).

Both understandings of space will be applied in the context of knowledge management (KM) and knowledge network management (KNM) in science and technology parks. Knowledge management measures and instruments are distinguished according to their mode of operation between relational (individual–individual) and topical (individual–space–(individual)) relationships. In particular, the latter opens up a new perspective on genuine spatial properties of KNM.

As a broad concept, knowledge management originates from the business management literature. In the scientific debate, it is referred to as the whole life cycle of knowledge processing, that is knowledge creation, validation, presentation, distribution, sharing and application (Bhatt 2001, Gloet and Terziovski 2004, Scholl et al. 2004). Active knowledge management practices are defined as organizational routines, as well as control and coordination mechanisms that organizations use intentionally to govern knowledge processes and to influence knowledge management outcomes; respectively in order to create or sustain competitive advantages (Quintas, Lefrere, and Jones 1997, Meier 2011, Scholl et al. 2004). Knowledge management varies in terms of systematization, normalization, complexity and relevance in the decision-making process. Today, it is generally understood as an organizational challenge to coordinate individual knowledge carriers, apply knowledge in business processes and to utilize technical support instruments as facilitators (Bullinger, Wörner and Prieto 1998). Several scholars identify two different KM strategies: codification, that is knowledge management based on IT solutions, through which knowledge is codified, stored and made accessible to anyone; and personalization, which centres on the promotion of direct personal interaction as knowledge is strongly linked to the individuals that have created it (Revilla, Acosta and Sarkis 2005).

At the beginning of the twenty-first century, a shift from technical-oriented knowledge management to personalized approaches was observed. Tactic knowledge and personal interaction have become the central focal point in the knowledge society (Meinke 2012). In today's Internet era, knowledge management is regarded as people management, which aims to connect people around the world based on ICT and the Internet as communication channels and virtual space of interaction (Henn and Meyhöfer 2003). Nevertheless, the direct links between individuals and organization are the only factors that matter. Personalized knowledge can also occur via direct interaction towards space by perceptive stimuli that specific environments offer. In order to bring people together, 'protected spaces' (*ba*) and 'enabling spaces' (Peschl and Fundneider 2012), like innovation labs or co-working spaces, are increasingly being established where individuals establish relationships towards a space first, independently of alternating co-located professionals and flexible superstructures (Suwala 2013; Lewis and Moultrie 2005, Magadley and Birdi 2009).

The literature predominantly emphasizes knowledge management at the firm level (for example Toyota) (Dyer and Nobeoka 2000) as well as in the inter-organizational context of strategic alliances, joint ventures and joint research teams (for example Xerox and Hitachi) (Cavusgil et al. 2003, Quintas, Lefrere and Jones 1997, Revilla, Acosta and Sarkis 2005, Hautala 2011, Meier 2011).[2] Empirical studies show that firms are more effective in transferring and sharing knowledge, when they actively implement knowledge management practices. Knowledge management instruments studied in strategic alliances include advisory systems, alliance liaison offices, learning networks, training programmes, as well as mutual visits, plant tours, and on-site meetings (Meier 2011). Communities of practice, brainstorming camps, virtual teams, mentorship programmes, storytelling, talent mobility, job rotation, inter-divisional teams, information and electronic learning networks, delegation of responsibility and reducing organizational hierarchy are further mechanisms in 'learning organizations' (Lundvall 2006, Scholl et al. 2004, Seufert 2000).

While other strands of the academic debate focus on the nature of knowledge and the different institutional levels of intervention, we examine the contextual implications of knowledge network management on the spatial scale of specific 'localities of learning' (Scholl et al. 2004, Meinke 2012).

Knowledge Network Management in Territorial Innovation Systems

Typically, territorial innovation systems are multi-actor innovation networks, which are characterized by heterogeneous entities and groups (for example companies, HEI, non-university R&D institutions, public policy institutions) and predominantly weaker structures compared to intra-firm networks. Economic success and competitive advantages of regions in the knowledge economy as well as rapidly changing global trends and environments greatly depend on the regions' ability to create and process knowledge and quality of its innovation networks (Harmaakorpi and Melkas 2005). The debate on proximity as a precondition of collective learning shows that non-spatial natures of proximity, in particular social, cognitive and organizational proximity, are critical for promoting interactive learning across different regional 'learning organizations' (Boschma 2005, Brinkhoff, Suwala and Kulke 2012).[3]

However, studies on knowledge (network) management in territorial innovation systems still remain very limited. In the RIS debate, for example, Dahl and Pedersen (2003) and Saxenian (1994) highlight the supporting culture and infrastructure, including trade fairs, conferences, seminars, and social activities, which facilitates the building of personal relationships and informal links in Silicon Valley. These informal networks also encouraged the sharing of market

2 Meier (2011) provides a comprehensive overview and discussion of the key aspects of knowledge transfer (e.g. knowledge characteristics, partner characteristics, partner interaction and active knowledge management in strategic alliances.

3 See Knoben and Oerlemans (2006) for a comprehensive literature review of the interplay of the different dimensions of proximity.

and technical information. In contrast, cultures and settings in other regional innovation systems such as the Route 128 case rather neglected the formation of informal networks and sharing of knowledge. In further studies, Fukugawa (2006, 2010) observes a positive effect of such distinct institutions and services (for example incubation manager, human resource development services) on linkages of science park-based companies to research centres. In the example of the regional innovation system of Lahti, Harmaakorpi and Melkas (2005) describe the creation of tailor-made knowledge management instruments responding to the specific natures of knowledge prevalent in each stage of their SECI / *rye bread model* of knowledge creation (see Table 1.1). In their model, self-transcending knowledge, which is defined as tacit knowledge prior to its embodiment (for example sensing the presence of a certain potential), is added as another type of knowledge (in addition to codified (explicit) and personalized (tacit) knowledge) (Harmaakorpi and Melkas 2005). Interestingly, the model also refers to different types of the *ba* during knowledge transformation. The main objective of our analysis will be to show what KNM instruments are best to being applied to specific stages of *ba*. Hereby, in particular topical measures will play a key role in addition to more common relational instruments.

In this chapter, we define knowledge network management (KNM) as mechanisms to organize and steer the generation and transfer of knowledge across organizational boundaries of those knowledge organizations and innovation networks that define the dynamic innovation 'ecosystem' (triple helix) on the geographical scale of distinct 'localities of learning'. It aims at tapping information flows, structuring knowledge and connecting knowledge carriers in innovation hubs such as science and technology parks, but also in the region as a whole. Knowledge network management tools, for example, comprise virtual communication platforms, networking and matchmaking events, knowledge marketing, financial support schemes for transfer of talent and joint research projects as well as intermediary institutions (Know-Man 2011).

The business management literature on knowledge management highlights different classifications of KM instruments (for example formal and informal learning behaviour, direct and indirect practices). In this chapter, we propose a distinction between relational (that is person-based) and topical (that is space-based, *ba*) knowledge network management instruments and contexts, respectively. Relational settings are characterized by direct steering mechanisms that promote the creation, transfer and application of knowledge between different individuals (individual–individual). In contrast, topical knowledge network management mechanisms (individual–space–(individual)) stress the importance of an active *ba* as a physical or virtual space (or framework). The *ba* serves as a common environment among like-minded individuals ('peers') and organizations. Therefore, personal interaction stimulates the dissemination and sharing of information and knowledge. Interaction between individuals is indirect as the communication predominantly is directed towards the *ba* in general.

Table 1.1 Knowledge management instruments in Lahti regional innovation system based on SECI / *rye bread model*

Phases of 'ba' (SECI / *rye bread model*)	**Knowledge types**	**Knowledge creation / management instruments**
Visualisation/ imagination 'ba' (identifying hidden trends and potentials)	*Self-transcending à tacit*	*Forecasts, scenarios, expert-based statements / interviews; external expertise and perspectives (unprejudiced selection of experts)*
Socialisation/ originating 'ba' (creation of a trustful atmosphere between actors of the core process)	Tacit à tacit	Meetings and events for socializing and networking in a relaxed atmosphere in order to increase social cohesion in the innovation network, e.g. sport events, experience and study trips, team-building events. A thematic alignment of events with different target groups (for example clients, users) is feasible, too.
Externalisation/ interaction 'ba' (promotion of productive learning)	Tacit à explicit	Thematic seminars with communication and sharing of the newest information and knowledge. Brainstorming and idea-development platforms and meetings that are moderated and documented to facilitate collective learning. Interactive forums facilitate interaction beyond certain meetings.
Combination/cyber 'ba'	Explicit à explicit	Internet-based extranet/virtual platform for combination of explicit knowledge, which consists of project plans, minutes of meetings, research reports, action models and best practices. Channel for external information and knowledge to get to the network
Internalisation/ exercising 'ba'	Explicit *à* tacit	Thematic group education with strong emphasis on practical exercises (various stakeholder involved). Exchange of experts between organisations
Potentialisation/ futurisation 'ba'	*Explicit à self-transcending*	*Delphi techniques, forecasts,etc.*

Source: synthesis of Harmaakorpi and Melkas (2005) findings.

The *ba* was primarily examined in intra-firm knowledge creation and organizational learning at the company level (Nonaka 1991, 1994). Senoo et al. (2007) show that workplace formation actively contributes to knowledge creation by activating the *ba*. KM practices that aim to enhance the exchange of information and knowledge as well as direct communication comprise workplace layout reformation (creation of free-seating systems, meeting areas and demo spaces boosting interaction and communication) and the development of an intranet database, in which internal documents were digitized systematically (Senoo et al. 2007).

In this section, we discuss how the model of an active *ba* can be applied to and created in science parks and other types of territorial innovation systems. Harmaakorpi and Melkas (2005) argue that modern firm organizations are mainly characterized as networked firm entities, and consequently, can be compared to territorial innovations systems. Differences in terms of hierarchy, leadership and general management, however, still exist (Harmaakorpi and Melkas 2005).

Table 1.2 illustrates the distinction of relational and topical KNM measures both enhancing the *ba* by using mechanisms mentioned in the knowledge management and regional innovation systems debate (Harmaakorpi and Melkas 2005, Meier 2011, Lundvall 2006, Scholl et al. 2004, Uotila, Merkas and Harmaakorpi 2005). Whereas knowledge management practices at the firm level mainly put an emphasis on the promotion of direct personal interaction, we argue that KNM should also stress topical modes of operation. However, in many cases topical instruments cannot be allocated explicitly to only one of the two categories, but show central features of both modes of operation simultaneously. In the case of a 'liaison office', its role is either to offer a 'protected space' (topical/*ba*) for inter-organization interaction or to facilitate direct relationships (relational). These circumstances can only be satisfactorily answered by qualitative methods. Therefore, we conducted in-depth interviews with businesses to assess the 'liaison office's function – either as 'safe haven' to detect suitable cooperation partners or its matchmaking process.

Managing Innovation in Two Science Parks in Berlin and Seville

The empirical framework of this chapter is examined at the two science parks Berlin-Adlershof [4] and Seville-Cartuja.[5] The data was gathered within the EU

---

4 The Berlin-Adlershof science park was established in 1991. Today ca. 955 companies (of which 445 high-technology companies primarily operate in the six key clusters optics and photonics, material and micro system technologies, ICT, environmental technologies, energy and biotechnology), 11 non–university research institutions and six natural science departments of the Humboldt-Universität zu Berlin are located at the site. In total, ca. 14,940 employees work in the science park – about 12,120 of them in the private sector and ca. 2,820 at the university and non–university research institutions. Furthermore, roughly 8,440 students study in the science park. In 2012, all organizations generated a total of ca. €1.7 billion in turnover (WISTA–MG 2013).

5 The Seville-Cartuja science park was founded in 1993 at the former site of the Expo 1992. In 2010, about 340 companies with a total of ca. 15,100 employees have been located there. About 180 of them are high–technology companies with about 7,860 employees that work in the science park's major clusters ICT, engineering, environmental technologies, biotechnology/agro-food, energy and health. Furthermore, five universities (e.g. the University of Seville and EOI Business School) with ca. 8,000 students and multiple R&D

Table 1.2 Relational and topical instruments of knowledge network management

Relational instruments	Topical instruments
Business management literature: • Alliance liaison offices • Learning networks • Information and electronic learning networks • Training programmes • Mentorship programmes • Advisory systems • Inter-divisional teams • Virtual teams • Talent mobility / job rotation • Storytelling • Mutual visits, plant tours, and on-site meetings RIS literature: • Thematic group education with strong emphasis on practical exercises • Incubation manager • Human resource services • Matchmaking events • Financial support schemes for joint research projects	Business management literature: • Communities of practice • Brainstorming camps RIS literature: • Socializing and networking meetings and events, e.g. sport events, experience and study trips, team-building events • Thematic seminars with communication and sharing of the newest information and knowledge • Knowledge marketing • Intermediary institutions (e.g. TTO) • Internet-based extranet / virtual communication platform for sharing and combination of explicit knowledge

Source: own compilation based on: Scholl et al. (2004), Melkas (2005), Fukugawa (2006, 2010), Lundvall (2006), Know-Man (2011), Meier (2011).

Interreg IVC project 'Knowledge network management in technology parks (Know-Man)' in Autumn 2010 and Spring 2011. In total, 54 technology-based firms (TBF) participated in a survey, which was characterized by a methodological mix of standardized questionnaires and expert interviews. In the emprirical research, we aim to analyse the character of interaction between high-technology businesses and scientific institutions as well as the impact of existing knowledge network management systems at the two innovation sites in Berlin and Seville.

The large majority of the surveyed firms are characterized as R&D intensive high-technology companies, which maintain strong or even multi-faceted long-term relationships to scientific institutions in the science parks and beyond.

Both science parks are large-scale regional redevelopment landmark projects, which aim to promote innovation and regional economic growth in distinct high-technology sectors. The triple helix, that is complementary HEI and

centres reside in the science park. In 2010, the organizations' total turnover was ca. €1.9 billion (Brinkhoff, Suwala and Kulke 2012).

non-university research institutions in conjunction with TBF, is a critical element at both science parks. The two case studies only differ slightly in terms of the businesses' composition. While the Berlin-Adlershof science park is characterized by a large number of micro and small companies, Seville-Cartuja's is also home to several individual large national companies and units of prominent international firms, for example, Alter Technology Group and Celgene (Brinkhoff, Suwala and Kulke 2012). At the beginning of both science parks' development, geographical proximity was the primary initial precondition for the development of networks and a dynamic growth process. Management strategies did not include any innovation network building systems (Kulke 2008, González Romero 2002, Brinkhoff, Suwala and Kulke 2012). However, various boundary-spanning mechanisms and KNM practices have been installed in the recent past. Figures 1.2 and 1.3 provide a chronological overview of important knowledge network management instruments and institutions in both science parks (bold letters) and related regions.

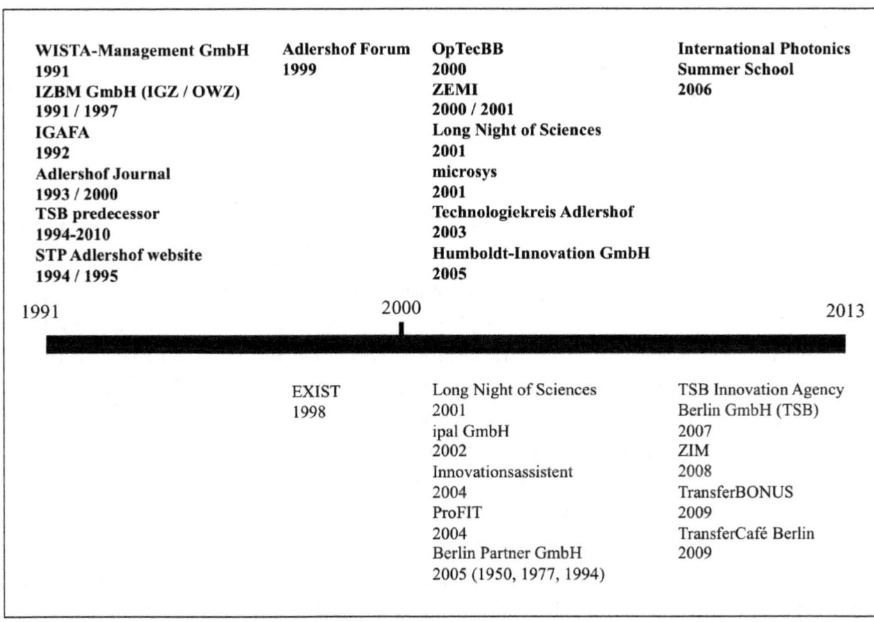

Figure 1.2 Timeline of major Berlin-Adlershof science park and Berlin third stream initiatives and KNM instruments

Source: authors' elaboration.

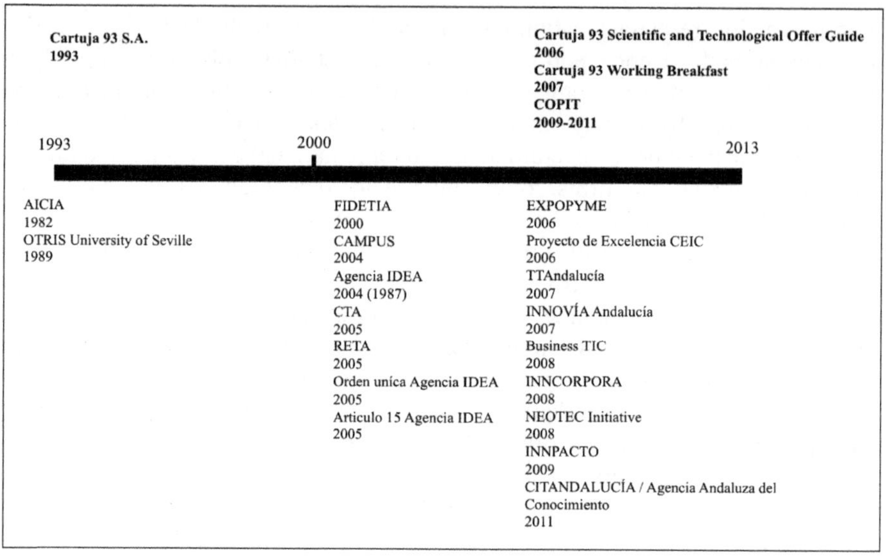

Figure 1.3 Timeline of major Seville-Cartuja science park and Andalusia third stream initiatives and KNM instruments
Source: authors' elaboration.

In particular, the qualitative case study approach allows the identification of a diverse set of substantial elements (for examples locational factors),[6] and – most importantly for this chapter – relational (individual–individual) and topical (individual–space–(individual)) KNM instruments that contribute to collective learning processes in the two science parks. Table 1.3 names the most prominent relational and topical KNM practices that are implemented in the Berlin-Adlershof and Seville-Cartuja science parks (as well as on the regional and national scale).

Previous research in inter-organizational relationships and innovation-related collaborative activities highlights the central position of personal contacts (Oinas 1997, Boschma 2005, Brinkhoff, Suwala and Kulke 2012). Typically, these ties are based on personal relationships to former work colleagues or fellow students. Other contacts are based on referrals from the companies or the scientific institutions' personnel. Also, our two case studies reveal the very strong importance of relational KNM practices and institutions that initiate personal interaction directly and, hence, the building of social proximity and mutual

6 As an illustration, we provide an example of the static, substantial and non–personalised element of KNM. Locational factors such as a critical mass of knowledge carriers are illustrated in the following statement collected in the Berlin–Adlershof case study: 'Although nothing has developed so far, here is a certain potential, which you won't find to this extent at another location'. (Company 18, STP Adlershof)

Table 1.3 Relational and topical KNM instruments at the two case studies

Type of KNM practices	Berlin-Adlershof science park	Seville-Cartuja science park
Relational knowledge network management mechanisms	• STP management WISTA-Management • Incubator management IZBM • Technology transfer offices / intermediaries (Humboldt-Innovation) • Industry and regional networks (e.g. Technologiekreis Adlershof, OpTecBB) • Matchmaking events • *Virtual contact platform Berlin TransferCafé* • *Talent-oriented support programmes (e.g. Innovationsassistent)* • *Financial support programmes for spin-offs and technology start-ups (e.g. EXIST)* • *Financial support programmes for R&D projects (e.g. ProFIT, TransferBONUS)*	• STP & incubator management Cartuja 93 S.A. • *Matchmaking events (e.g. COPIT, encuentros TT)* • *Technology transfer offices / intermediaries (e.g. FIDETIA, CITANDALUCÍA, CTA, TTO of University of Seville)* • *Regional development and innovation agency IDEA* • *Talent-oriented support programmes (e.g. INNCORPORA, Torres Quevedo)* • *Financial support programmes for spin-offs and technology start-ups (e.g. Programa CAMPUS, Articulo 15 Agencia IDEA, NEOTEC)* • *Financial support programmes for R&D projects (e.g. Orden única Agencia IDEA, INNPACTO, Proyecto de Excelencia CEIC)*
Topical knowledge network management practices	• R&D marketing (e.g. WISTA website, Adlershof Journal) • Networking events (e.g. Forum Adlershof Business Lunch) • Congresses / conferences (e.g. microsys Berlin) • International Photonics Summer School • Long Night of Sciences	• R&D marketing (e.g. Cartuja 93 website, Cartuja Scientific Inventory) • Networking events (e.g. Cartuja 93 Working Breakfasts, COPIT) • *Congresses / conferences (e.g. Expo PYME, Innovia, Business TIC)*

Note: KNM instruments *in italic* letters are implemented beyond the science park-scale – on the regional or/and national scale.
Source: authors' elaboration.

trust. Intermediary institutions at both innovation sites such as the science park management organizations WISTA-Management GmbH and Cartuja 93 S.A. hold an active matchmaking role. Further boundary-spanning interfaces are the locally rooted networks Technologiekreis Adlershof and OpTecBB in the Berlin science park, as well as distinct Andalusian technology transfer organizations, for example CTA and FIDETIA. The following statements showcase the emphasis on direct relations between different knowledge organizations (key words are highlighted in italic letters).

> The Technologiekreis [Adlershof] is a very important institution. ... Often, requests for *cooperation* are stated. You just have to know whether it suits your interests and whether you have the personnel to realize it within the time period. Then, you can *get in touch with the people*. (Company 13, STP Adlershof)

> It's always easier, when you know each other beforehand, when you know who you are talking to compared to when you have to ask through the university in order to *find the right person*. ... This is the reason why we are in various committees and networks such as OpTecBB. Not because we're expecting big business, but to get to *know* beforehand *who* works in what field and who could contribute to solutions in certain problems at some point. (Company 9, STP Adlershof)

In these two examples, the Berlin-based businesses' CEOs know exactly whom they are going to meet through the networks Technologiekreis Adlershof and OpTecBB. The network members know each other and also are known to prospective members (through the networks' marketing activities). Know-who is the crucial circumstance for relational KNM instruments. Hence, the businesses already presume what kind of relationships with distinct businesses and other organizations may evolve through their active involvement in the business networks.

Nevertheless, the interviewees also emphasize topical relationships that are manifested in different *bas*. Exemplary modes are virtual or physical spaces through which individuals and organizations connect indirectly (socializing ba), but subsequently may result in direct personal links. These anchoring spaces may be theme-marked or experience-marked places or distinct networking events, meetings points and virtual platforms that 'peers' utilize and frequent.

> Most importantly, the framework is provided by the informal contact, for example the informal contact when you get to know each other (…) and when it becomes transparent what they others can do. (…) There's the 'Academic Lunch', *where* one and the other CEO are invited and sit next to a university professor at the table. And as soon as three people sit together, they talk 'what do you do?' and 'what do you do?' and such things. (Company 19, STP Adlershof)

> Fundamentally, it's been through personal contacts, but always under the umbrella of Cartuja 93 S.A., the management entity of the STP Cartuja, with distinct activities, events, conferences, talks, and so on. Well, *there* you will meet research groups, and so on, you will exchange ideas and opinions. (Company 22, STP Cartuja)

> Contacts are made at conferences, trade shows and presentations. These are good *occasions and places*, respectively, to *meet people* that work in the same field or something which may create synergies. (Company 14, STP Cartuja)

Furthermore, multiple interviewees express the demand for rather topical KNM practices such as physical places and events (physical *ba*). As an illustration, science park-based demo shows at prominent and highly frequented places such as the Cartuja science park's business centre Pabellón de Italia could create a *ba*, which spurs informal interaction and the building personal relationships.

> But what I'm missing is *some space where* demos, prototypes, products, and so on, of the companies are presented. (...) We can pass it, we have a look at the demo, and can go from there to get in touch in a natural way. (...) More informal, more non-virtual. (Company 10, STP Cartuja)

> I would like them to meet me. To *host a conference and a forum*, and the research centres tell us about what they focus on, what they can offer, what they dedicate to, where do they want to be, how do they think they could support the companies ... I think it could be done more quickly, to sit down together at a conference, a forum for one day, and to tell us. (Company 25, STP Cartuja)

> One way to build contacts would be *a regular event*, for example once a week. ... Fun and a relaxed atmosphere should be underlined, probably at a morning pint or brunch. (Company 6, STP Adlershof)

'Know-where', the physical *ba*, plays the crucial role here. Moreover, a virtual *ba* in terms of R&D marketing and information instruments such as comprehensive and demand-oriented virtual information and communication platforms are underlined by some companies, in particular those working in the ICT sector.

> There should be an *Internet platform, where* companies could post requests for student placements (for example internships, master theses, working groups) for all universities. (Company 3, STP Adlershof)

> ... also, some kind of online database, a *platform* through which the companies can get in touch ... And it's also about the daily business, the products, the news, the ideas they have ... But you can put something on it like 'I'm looking for a technology partner for project X'. (Company 28, STP Cartuja)

Virtual spaces (for example social media platforms) stress the 'knowing where' and the indirect interactive relationships between individuals and organizations through 'protected spaces' (*ba*). Finally, intermediary institutions and persons take the role as personalized or institutionalized *ba*. Here, we interpret the role of such intermediary institutions as 'space', which can be considered as an intermediary membrane. There, the *ba* receives, transports and disseminates messages and needs between different individuals and organizations. This leads us to institutions (for example boundary spanners and networks) that bridge space through a so-called institutional *ba*.

> I think a necessary service *is a [organization] that facilitates* communication between the scientific and the business world. This service can be implemented by *active officers* that promote cooperation between both sides, for example, innovation promoters. (Company 2, STP Cartuja)

> If there would be a *care-taker* that is associated to WISTA or something, that promotes the organization in the network, it would be great. You could place your requests, your ideas to him and he would try to realize them. (Company 24, STP Adlershof)

However – as stated before – the role of brokerage institutions such as the science park management and industry networks is not clearly assigned to either relational or topical modes (institutional *ba)*. These agents rather embody a hybrid function. The following example illustrates this aspect.

> About 50 per cent of the connections came from the contacts I had from my *previous research work experiences.* The other 50 per cent of the relationships came *through contacts to CSIC* or are *referred by the CTA*, for example. Also, through a networking event initiated by the research group of CABIMER – they facilitated the meeting of companies and researchers. (Company 12, STP Cartuja)

In many cases, the allocation of KNM practices to either one of the categories remains ambiguous. Relational and topical modes mostly coexist parallel with their respective individual features often complementing each other. Ultimately, knowledge network management is people management. Topical modes, that is institutional *ba*, virtual and physical *ba*, are created spaces that attract cognitively proximate knowledge carriers ('peers') and enable and accelerate the building of inter-personal relationships. Almost all examples from the in-depth interviews that we illustrate in this chapter exclusively refer to the phase of 'socializing / originating *ba*' in Nonaka's SECI model (which also is included in the adapted *rye bread model* of Haarmokorpi and Melkas 2005).

Discussion and Conclusion

This chapter offers insights on knowledge network management in TIS as localities of learning. We define KNM as distinct practices and institutions to coordinate the creation and transfer of knowledge among knowledge organizations and innovation networks of the 'triple helix' in TIS. Our case studies at the Berlin-

Adlershof and Seville-Cartuja science parks illustrate that managing innovation and knowledge in science parks as 'localities of learning' requires a complementary set of both relational and topical instruments of KNM. Therefore, science parks have to move beyond being a simple geographical concentration of private companies, HEI and R&D institutions, and political institutions (as propagated in the industrial society). Businesses of the knowledge economy clearly expect intermediary institutions such as science parks to provide direct and indirect links to co-located potential partner organizations in the private sector and academia. However, geographical proximity solely is not a sufficient precondition. Also, additional natures of proximity, namely organizational proximity in terms of active KNM systems, have to be implemented to nurture the development of inter-organizational linkages (Brinkhoff, Suwala and Kulke 2012). Consequently, modern science parks have to evolve towards active knowledge-creating entities. Taking this boundary-spanning role, innovation hubs have to create different *bas* in inter-organizational relationships and networks among stakeholders in the regional 'triple helix' (Hansson 2007). The findings of our case studies show that in particular topical KNM instruments represent genuine space-dependent measures to manage knowledge in territorial innovation systems through direct manifold individual–space (for example physical, virtual and institutionalized) relationships. These results are further supported by empirical studies at the firm level by Alvarenga Neto and others (Alvarenga Neto 2005, 2008; Alvarenga Neto et al 2009). Their findings of the assessment of KM initiatives in over 20 international firms (for example 3M, Dow Chemical, Xerox, Ernest & Young, British Telecom and Mircosoft) suggest that the creation of *ba* or similar enabling conditions and contexts are not only crucial when managing knowledge in networks; knowledge organizations rather manage the context and readiness where knowledge is socially created, shared and utilized (Choo and Alvarenga Neto 2010).

Nevertheless, we support today's understanding of knowledge network management as (a type of) people management in particular. However, on the spatial dimension of TIS and science parks, the *ba*, in terms of 'protected enabling spaces' becomes more important in order to bring the multi-faceted types of knowledge carriers, organizations and networks together. These mutual spaces are defined by distinct needs, expectations, experiences, themes and knowledge backgrounds. Their role, which is recently emphasized by the increasing academic interest in organizational innovation spaces, for example innovation labs and co-working spaces (Lewis and Moultrie 2005; Magadley and Birdi 2009), becomes more crucial as they serve as initializing conglomeration for interactive relationships that contribute to the innovativeness of TIS and 'localities of learning'.

Acknowledgements

The study has benefited from the EU Interreg IVC project Knowledge Network Management in Technology Parks (Know-Man), which was implemented from

2010 to 2012. Further information is available at the project's website (www.know-man.eu).

References

Alvarenga Neto, R. 2005. *Gestão do conhecimento em organizações: proposta de mapeamento conceitual integrativo* [Knowledge management in Organizations: An Integrative Conceptual Mapping Proposition]. Belo Horizonte: Doctoral thesis in Information Science (UFMG).

Alvarenga Neto, R. 2008. *Gestão do conhecimento em organizações: proposta de mapeamento conceitual integrativo* [Knowledge Management in Organizations: An Integrative Conceptual Mapping Proposition]. São Paulo: Editora Saraiva.

Alvarenga Neto, R., Souza, R.R., Queiroz, J.G. and Chipp, H. 2009. *Implementation of a Knowledge Management Process within the Brazilian Organizational Context: the Ons (National Operator of the Interconnected Power System) Experience*. Paper to the 6th International Conference on Intellectual Capital, Knowledge Management and Organisational Learning, School of Information Studies, McGill University, Montreal, 1–2 October 2009, London, Academic Conferences International.

Asheim, B.T. 1997. Learning regions in a globalised world economy: Towards a new competitive advantage of industrial districts?, in *Interdependent and Uneven Development: Global–Local Perspectives*, edited by S. Conti and M. Taylor. Aldershot: Ashgate, 143–76.

Aydalot, P. 1986. *Milieux Innovateurs en Europe*. Paris: GREMI.

Bhatt, G.D. 2001. Knowledge management in organizations: examining the interaction between technologies, techniques, and people. *Journal of Knowledge Management*, 5(1), 68–75.

Boschma, R.A. 2005. Proximity and innovation: a critical assessment. *Regional Studies*, 39 (1), 61–74.

Bouchara, M. 1987. L'industrialisation rampante: ampleur, mécanismes et portée. *Economie et Humanisme*, 297, 37–49.

Braczyk, H.-J., Cooke, P. and Heidenreich, M. (ed.) 1998. *Regional Innovation Systems*. London: UCL Press.

Brinkhoff, S., Suwala, L., and Kulke, E. 2012. 'What do you offer?': Interlinkages of universities and high-technology companies in science and technology parks in Berlin and Seville, in *Universities, Cities and Regions. Loci for Knowledge and Innovation Creation,* edited by R. Capello, A. Olechnicka and G. Gorzelak. London, New York: Routledge, 121–46.

Bullinger, H.-J., Wörner, K. and Prieto, J. 1998. Wissensmanagement. Modelle und Strategien für die Praxis, in *Wissensmanagement. Schritte zum intelligenten Unternehmen,* edited by H.D. Bürgel. Berlin: Springer, 21–39.

Cavusgil, S.T, Calantone, R.J. and Zhao, Y. 2003. Tacit knowledge transfer and firm innovation capability. *Journal of Business & Industrial Marketing*, 18(1), 6–21.

Choo, C.W. and Alvarenga Neto, R. 2010. Beyond the ba: managing enabling contexts in knowledge organizations. *Journal of Knowledge Management*, 14(4), 592–610.

Couclelis, H. 1992. Location, place, region and space, in *Geograph's Inner Worlds*, edited by R.F. Abler, M.G. Marcus and J.M. Olson. New Jersey: Rutgers University Press, 215–33.

Dahl, M.S. and Pedersen, C.O.R. 2003. *Knowledge Flows through Informal Contacts in Industrial Clusters: Myths or Realities?*, DRUID Working Paper No. 03–01, Aalborg.

Dosi, G. 1996. The contribution of economic theory to the understanding of a knowledge-based economy, in *Employment and Growth in the Knowledge-based Economy*, edited by OECD. Paris: OECD, 81–93.

Dyer, J.H. and Nobeoka, K. 2000. Creating and managing a high-performance knowledge-sharing network: the Toyota case. *Strategic Management Journal*, 21(3), 345–67.

Etzkowitz, H. and Leydesdorff, L. (eds) 1997. *Universities in the Global Economy: A Triple Helix of University–Industry–Government Relations*, London: Cassell Academic.

Etzkowitz, H. and Leydesdorff, L. 2000. The dynamics of innovation: from national systems and "Mode 2" to a triple helix of university–industry–government relations. *Research Policy*, 29(2), 109–23.

Florida, R. 1995. Toward the Learning Region. *Futures*, 27(5), 527–36.

Fukugawa, N. 2006. Science parks in Japan and their value-added contributions to new technology-based firms. *International Journal of Industrial Organization*, 24 (2), 381–400.

Fukugawa, N. 2010. *Assessing the Impact of Science Parks on Knowledge Interaction in the Regional Innovation System*, Summer Conference 2010 Opening Up Innovation: Strategy, Organization and Technology, Imperial College London Business School, 16–18 June, 2010.

Gloet, M. and Terziovski, M. 2004. Exploring the relationship between knowledge management practices and innovation performance. *Journal of Manufacturing Technology Management*, 15(5), 402–9.

González Romero, G. 2002. Actividad Innovadora y Creación de Sinergias en el Complejo Innovador de Sevilla-Technópolis. *Archivo Hispalense*: 259–60.

Hansson, F. 2007. Science parks as knowledge organizations – the 'ba' in action? *European Journal of Innovation Management*, 10(3), 348–66

Harmaakorpi, V. and Melkas, H. 2005. Knowledge management in regional innovation networks: the case of Lahti, Finland. *European Planning Studies*, 13(5), 641–60.

Hautala, J. 2011. International academic knowledge creation and ba. A case study from Finland. *Knowledge Management Research & Practice*, 9(1), 4–16.

Henn, G. and Meyhöfer, D. 2003. *Architecture of Knowledge*. Oxford: Butterworth-Heinemann & Architectural Press.
Huber, F. 2012. Do clusters really matter for innovation practices in information technology? Questioning the significance of technological knowledge spillovers. *Journal of Economic Geography*, 12(1), 107–26.
Katai, O., Minamizono, K., Shiose, T. and Kawakami, H. 2007. System design of 'Ba'-like stages for improvisational acts via Leibnizian space–time and Peirce's existential graph concepts. *AI & Society*, 22(2), 101–12.
Knoben, J. and Oerlemans, L.A.G. 2006. Proximity and inter-organizational collaboration: a literature review. *International Journal of Management Reviews*, 8(2), 71–89.
Know-Man 2011. *Know-Man Good Practices. Knowledge Network Management in Technology Parks*. Warsaw: Wydawnictwo Naukowe Scholar.
Kujath, H.J. and Stein, A. 2011. Lokale Wissensbeziehungen in den globalen Beziehungsräumen der Wissensökonomie, in *Räume der Wissensarbeit. Zur Funktion von Nähe und Distanz in der Wissensökonomie*, edited by O. Ibert and H.J. Kujath. Wiesbaden: VS Verlag für Sozialwissenschaften, 127–54.
Kulke, E. 2008. The technology park Berlin-Adlershof as an example of spatial proximity in regional economic policy. *Zeitschrift für Wirtschaftsgeographie*, 52(4), 193–208.
Latka, T. 2003. *Topisches Sozialsystem. Die Einführung der japanischen Lehre vom Ort in die Systemtheorie und deren Konsequenzen für die Theorie der sozialen Systeme*. Heidelberg: Carl-Auer-Systeme Verlag.
Lewis, M. and Moultrie, J. 2005. The organizational innovation laboratory. *Creativity and Innovation Management*, 14(1), 73–83.
Lundvall, B.A. 1988. Innovation as an interactive process: from user–producer interaction to the national system of innovation, in *Technical Change and Economic Theory*, edited by G. Dosi et al. London: Pinter, 349–69.
Lundvall, B.A. (ed.) 1992. *National Systems of Innovation: Towards a Theory of Innovation and Interactive Learning*. London and New York: Frances Pinter.
Lundvall, B.A. 2006. *Knowledge Management in the Learning Economy*, DRUID Working Paper No. 06–6, Aalborg.
Magadley, W. and Birdi, K. 2009. Innovation labs: an examination into the use of physical Spaces to Enhance Organizational Creativity. *Creativity and Innovation Management*, 18(4), 315–25.
Markusen, A. 1996. Sticky places in slippery space. A typology of industrial districts. *Economic Geography*, 72 (3), 293–313.
Meier, M. 2011. Knowledge management in strategic alliances: a review of empirical evidence. *International Journal of Management Reviews*, 13, 1–23.
Meinke, J.H. 2012. *Wissensmanagement im Bereich der universitären Forschung. Ergebnisse einer Delphi-Studie im Hochschulbereich*. Regensburg: Doctoral thesis, Regensburg University.
Morgan, K. 1997. The learning region: institutions, innovation and regional renewal. *Regional Studies*, 31(5), 491–503.

Moulaert, F. and Sekia, F. 2003. Territorial innovation models: a critical survey. *Regional Studies*, 37(3), 289–302.

Nishida, K. 1970. *Fundamental Problems of Philosophy*. Tokyo: Sophia University.

Nishida, K. 1999. *Logik des Ortes* (translated to German and edited by R. Elberfeld). Darmstadt: WBG.

Nonaka, I. 1991. The knowledge-creating company. *Harvard Business Review*, 69(6), 96–104.

Nonaka, I. 1994. A dynamic theory of organizational knowledge creation. *Organization Science*, 5(1), 14–37.

Nonaka, I. and Takeuchi, H. 1995. *The Knowledge-Creating Company. How Japanese Companies Create the Dynamics of Innovation*. Oxford: Oxford University Press.

Nonaka, I. and Konno, N. 1998. The concept of 'ba': Building a foundation for knowledge creation. *California Management Review*, 40, 40–54.

Oinas, P. 1997. On the socio-spatial embeddedness of business firms. *Erdkunde*, 51(1), 23–32.

Peschl, M.F., Fundneider, T. 2012. Spaces enabling game-changing and sustaining innovations: Why space matters for knowledge creation and innovation. *Journal of Organisational Transformation and Social Change (OTSC)* 9(1), 41–61.

Pfister, D. 2007. *Raum – Gestaltung – Marketing im ganzheitlich-nachhaltigen Management*. Basel: Gesowip.

Porter, M. 1990. *The Competitive Advantage of Nations*. London: Macmillan.

Quintas, P., Lefrere, P. and Jones, G. 1997. Knowledge management: a strategic agenda. *Journal of Long Range Planning*, 30(3), 385–391.

Revilla, E., Acosta, J. and Sarkis, J. 2005. Value perceptions and performance of research joint ventures. An organizational learning perspective. *Journal of High Technology Management Research*, 16, 157–172.

Saxenian, A. 1994. *Regional Networks: Industrial Adaptation in Silicon Valley and Route 128*. Cambridge, MA: Harvard University Press.

Scholl, W., König, C., Meyer, B. and Heisig, P. 2004. The future of knowledge management: an international delphi study. *Journal of Knowledge Management*, 8(2), 19–35.

Senoo, D., Magnier-Watanabe, R. and Salmador, M.P. 2007. Workplace formation, active ba and knowledge creation. From a conceptual to a practical framework. *European Journal of Innovation Management*, 10(3), 296–315.

Seufert, S. 2000. *Work-Based Learning and Knowledge Management: An integrated Concept of Organizational Learning*, European Conference on Information Systems, Vienna, 3–5 July: 1413–20.

Shimizu, H. 1995. Ba-Principle: new logic for the real-time emergence of information. *Holonics*, 5(1), 67–79.

Siegel, D.S., Waldman, D. and Link, A.N. 2003. Assessing the impact of organizational practices on the productivity of university technology transfer offices: an exploratory study. *Research Policy*, 32 (1), 27–48.

Siegel, D.S., Westhead, P. and Wright, M. 2003. Assessing the impact of university science parks on research productivity: exploratory firm-level evidence from the United Kingdom. *International Journal of Industrial Organization*, 21(9), 1357–69.

Siegel, D.S., Waldman, D., Atwater, L. and Link, A.N. 2003. Commercial knowledge transfers from universities to firms: improving the effectiveness of University–Industry Collaboration. *The Journal of High Technology Management Research*, 14, 111–33.

Storper, M. and Scott, A.J. 1989. The geographical foundations and social regulation of flexible production complexes, in *The Power of Geography*, edited by J. Wolch and M. Dear. London: Allen & Unwin, 19–40.

Suwala, L. 2013. *Kreativität, Kultur und Raum – ein wirtschaftsgeographischer Beitrag am Beispiel des kulturellen Kreativitätsprozesses*. Wiesbaden: Springer VS.

Thune, T. 2009. Proximity and interactive learning in university–firm relationships. *Industry and Higher Education*, 23(1), 7–16.

Torre, A. and Rallet, A. 2005. Proximity and localization. *Regional Studies*, 39(1), 47–59.

Uotila, T., Melkas, H. and Harmaakorpi, V. 2005. Incorporating futures research into regional knowledge creation and management. *Futures*, 37(8), 849–66.

Vedovello, C. 1997. Science parks and university–industry interaction: geographical proximity between the agents as a driving force. *Technovation*, 17(9), 491–502.

WISTA-MG 2013. *Report on Adlershof*. Berlin: WISTA-Management GmbH.

Youtie, J. and Shapira, P. 2008. Building an innovation hub: a case study of the transformation of university roles in regional technological and economic development. *Research Policy*, 37, 1188–204.

Chapter 2
Industrial and R&D Activity in Emerging Economies – Poland versus other EU Countries

Tomasz Rachwał, Krzysztof Wiedermann

Introduction

The transformation of the socio-economic system in Poland and other emerging economies in Central and Eastern Europe, ongoing for over two decades, has contributed to profound changes in the region's economy, including the spatial organisation of industrial production. The economic structures formed under the centrally planned system underwent a gradual transformation, leading to greater spatial diversity on a regional scale, which has been of interest to many researchers (Gorzelak 1993, 2000, Szlachta 1993, Domański R. 1997, Kukliński, Mync and Szul 1997, Czyż 1998, Parysek 1998, Stryjakiewicz 1999, Domański B. 2001, Zioło 2006). These change processes have also been significantly influenced by changes in industrial activity, as reflected by the relocation and concentration of new enterprises in certain regions. In particular, this is indicated by foreign investment flowing into select regions. This may be accompanied by the closure of existing industrial plants (Pavlinek and Smith 1998, Domański B. 2001, Domański B. and Gwosdz 2005).

Consequently, it may be assumed that regional development in Poland after 1989 has been governed primarily by market mechanisms, and actions taken within the framework of economic policy aiming to compensate for differences in regional development, for instance by stimulating the process of the deconcentration of industry and economic stimulation in peripheral regions, have not yet brought about the expected results. In this situation, it appears important to answer the following two questions: Where in Europe are new manufacturing companies being established or expanded? Second, how will employment at manufacturing companies, industrial gross value added, and the value of investment change?

In the pursuit of a knowledge-based economy, industrial innovation is very important (Gierańczyk 2009a, 2009b, Gierańczyk 2010, Gierańczyk and Rachwał 2012, Rachwał 2012, Rachwał and Boguś 2012a, 2012b) as is its relationship with research and development, which is an important factor affecting regional development (Domański B. 2000, Hansen and Winther 2011). This is particularly important in the era of R&D globalisation (Gassmann and von Zedtwitz 1998, 1999).

The purpose of the chapter is to describe the relationship between business innovation and regional development as well as to identify EU regions with high-tech industries and those with a rapidly growing high-tech sector. The analysis of industrial change in the region's economic structure was based on the findings of previous studies on changes in employment, gross added value, productivity, and labour costs, while the consequences of the transformation of industrial structures were related to data on the value of investment in fixed assets, employment rates in the high-tech industry, R&D investment, and employment in this sector, which is widely considered to be a fundamental indicator of the level of innovation (Boix and Trullén 2010). The above listed indicators of industrial growth and structure were compared with economic development data in the form of GDP per capita (purchasing power parity). This comparison made it possible to show the relationship between industrial growth, especially growth in the high-tech sector, and economic development in different areas of the European Union.

Determining the changes in the value of these measures and indicators and their relationship to regional systems would lead to the identification of locations where new development benefits are accumulated, which is important from the perspective of the development of these areas. The role of new investment and the development of existing industrial firms in a region is significant, not only because of the direct impact of companies on local and regional labour markets, but also by generating multiplier effects in the economy (Garnick 1970, Wiedermann 2007, 2008, Domański and Gwosdz 2008, 2010, Rachwał and Wiedermann 2008, Gwosdz 2011, Micek 2011). The geographic scope of the study includes EU regions (NUTS 2) with a special focus on Polish provinces (voivodeships) and EU nations in terms of R&D investment data from the largest European corporations. However, multiplier effects of investment activity affect development across areas larger than a corporation's home region, which warrants country-level analysis.

The study area and the study period (1998–2010) were both largely determined by the availability of data. Data on regional systems are published later than data on countries, which is why often the latest regional data come from 2010. The analysis in the chapter is based on data from Eurostat and the Central Statistical Office of Poland (Polish acronym: GUS). It also takes into account the results of previous studies on innovation and the competitiveness of industry in Poland and other European countries (Gierańczyk 2003, 2008, 2009a, 2009b; 2010, Gierańczyk and Rachwał 2012, Rachwał and Boguś 2012a, 2012b), the branch structure of Polish industry (Rachwał 2010, 2011), the spatial diversity of industry in EU regional systems (Rachwał, Wiedermann and Kilar 2008a, 2008b, Rachwał, Wiedermann and Kilar 2009) as well as the spatial adaptation (Stryjakiewicz 1999) and changes in Polish industry versus that in Eastern and Central Europe (Domański 2003, 2006; Rachwał 2009). The analysis refers to the theoretical model of the operation of geographic space developed by Zioło (1996, 1999, 2009), in which there are certain patterns and relationships between its elements, including those between the basic elements of the spatial structure of industry (Zioło 2003).

High-Tech Industry in Regional Development – A Theoretical Approach

The problem of industrial restructuring relative to urban and regional economics is one of the key economic challenges in contemporary social and economic development. This subject has been frequently analysed with respect to plant relocation, which has occurred quite often in the world's most advanced economies. On the other hand, certain new regions have become attractive to certain types of business activity, which has produced new local and regional industrial structures. As a consequence, it has become necessary to identify new regional development paths, which have garnered the attention of researchers in the fields of human geography, regional economics, urban sociology, and community sociology. Key research papers in this area include those on 'learning regions', especially in the context of industrial areas (Asheim 1996), and papers on competitiveness (Amin and Thrift 1995, Storper 1997), which leads to the formation of industrial and innovation clusters (Porter 1990, 1998, 2008).

In light of strong global competition, advanced economies must focus on innovation in order to remain competitive. This is the view of Michael Porter (2003) who argued that the performance of regional economies is strongly influenced by the strength of local clusters and the vitality and plurality of innovation. This suggests a change in the economic base of a region, with an ever increasing role played by regional and local innovation systems (Malecki and Chesire 2004). Hence, it may be inferred that business innovation is the key to the creation of a modern economy that makes it possible for regions to grow (Figure 2.1).

An innovative company that responds quickly to market needs requires a highly qualified workforce, which includes R&D staff. Other key characteristics include significant R&D expenditures, acquisition of fixed assets, investment permanence, and embeddedness in the local and regional economy. These factors of regional development help sustain it over longer periods of time. In Central and Eastern Europe, new industrial investment often does not revolve around research and development, but is designed to seek out a low cost labour force. In this case, companies do not invest in R&D and are poorly embedded in the local economy (Grabher 1994). Mattes (2012) also notes this issue, asserting that the regional embeddedness of innovation projects of multinational companies does not necessarily have to occur even if a given region offers a wide array of collaborative opportunities and linkages at the level of regional institutions. The only local outcome of the presence of a company may be a small improvement in the local labour market due to the labour-intensive nature of work at a given facility. Pay is also relatively low at poorly embedded companies and their investment is not very permanent. When a region experiences an increase in economic development, pay tends to rise, and poorly embedded companies shutter their operations and migrate to areas with lower labour costs. However, this is not often the case with manufacturing companies pursuing research and development in a given region.

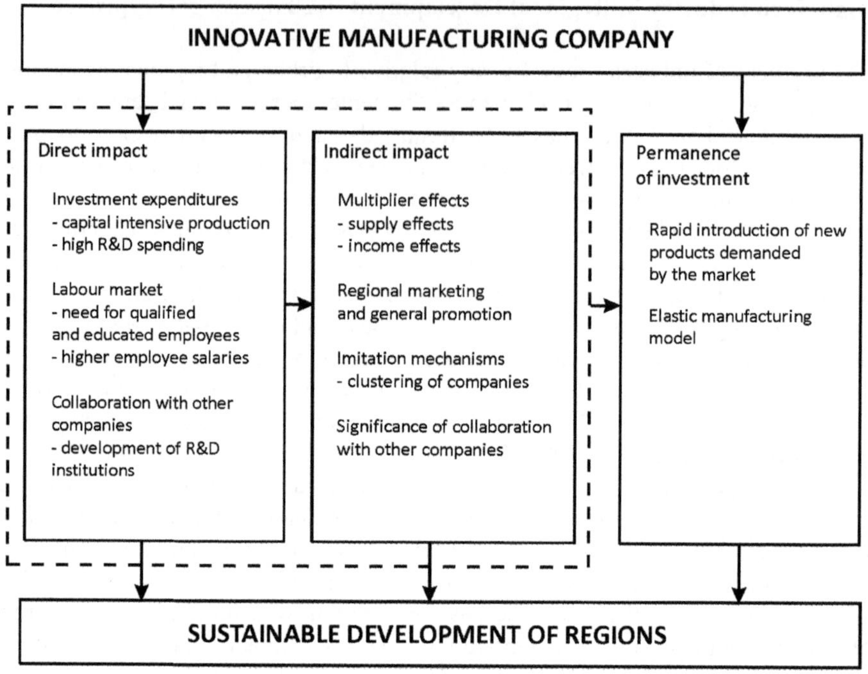

Figure 2.1 The business innovation – regional sustainable development relationship model

Source: authors' elaboration.

The Role of Industry in the Economy in Terms of Employment and Gross Value Added

Increasing levels of innovation at manufacturing companies and high-tech companies effect changes in the role of industry in the local economy, which becomes increasingly less reliant on large numbers of workers.

Earlier research on changes in employment in industry in regional systems suggests a declining role of industry in the labour market. The reduction in employment in industry mainly refers to countries in Central and Eastern Europe, where industrialisation in the period of the planned economy led to a large surplus of jobs in industry in relation to other types of economic activity. In the course of the region's recent economic transformation, the processes shaping the labour market in the region relied primarily on increasing the supply of jobs in the service sector.

The situation is somewhat different in the case of gross value added in industry. At current prices, this value increased in virtually all the regions of Europe.

Particularly high rates were observed in some regions of Central and Eastern Europe as well as Ireland and Portugal. In Polish voivodeships (provinces), this growth was also relatively high. In a few regions of Europe, there was a decrease in gross value added in industry. It should be taken into account, however, that this value also increased in other sectors of the economy, especially in services, which often stemmed from the fact that, despite high growth in value added, industry was losing its share in total gross value added. As a result of changes in employment growth in industry, its share in total employment also changed. Still, there is a large share of industry in total employment in many regions of Central and Eastern Europe, including the Polish southwest, and the traditionally highly industrialised regions of Germany, Spain and northern Italy. However, it is worth noting that the share was much higher in the case of gross value added (GVA) – more than 30 per cent in many parts of Europe. These indicators were higher in most regions of Europe than the share of industry in employment. The large share of industry in GVA was characteristic not only for the regions of Central and Eastern Europe, but also for the regions of Ireland, Finland, southern Germany, eastern Spain and northern Italy. Calculating gross value added per one employee allows for an analysis of workforce productivity in industry. In the case of this indicator, the most significant was the dividing line between former socialist countries and the most developed regions in Western and Northern Europe. In addition to workforce productivity, the second element of the competitive labour market in terms of attracting industrial activity is the cost of labour. In this case, the cost of a man-hour in the processing industry in Central and Eastern Europe is significantly lower. This facilitates the process of relocating production characterised by higher labour intensity to those regions. In this case, the difference in labour costs exerts the greatest impact on improving the competitiveness of production. Moving production, thus, most often concerns the least complex products, that are those located at the lower levels of supply chains.

What is crucial for industrial activity, from the standpoint of competitiveness of regions, is the analysis of the relationship between workforce productivity and average monthly labour costs in manufacturing. In this case, it is clear that the regions of Central and Eastern Europe, with generally lower labour costs, are not the most competitive in the European Union, especially in the area of non-labour-intensive industrial activity. In this respect, the most favourable indicators are observed in Spain, Ireland, southern Germany, Benelux and the Scandinavian countries. The analysis of the spatial differentiation of competitiveness of Polish regions in this area justifies, in economic terms, the currently observed inflow of industrial investment, which shifts towards the most developed industrial cities of southwestern and central Poland. For high value added production, for which productivity indicators are key factors, the most competitive regions are Opolskie, Silesian (Śląskie), Mazovian (Mazowieckie) and Lower Silesian (Dolnośląskie) Provinces. For more labour-intensive production, where adequate workforce productivity at lower cost is a more important factor, Łódzkie Voivodeship is the most competitive region in Poland (Rachwał, Wiedermann and Kilar 2008b, Rachwał and Boguś 2012a, 2012b).

Investment in Fixed Assets in Industry as a Sign of Competitiveness

The following question may be posed: How important is competition for new industrial investment in terms of the characteristics of regional labour markets? There are other important, if not key, factors such as accessibility, economic policy as well as the engagement of local and regional authorities. One of the indicators showing the influx of new businesses or the development of industrial activity is investment in fixed assets in industry. Putting them together with previous data on the competitiveness of regions may reveal the extent to which this affects the growth of the value of investment in industry. The greatest influx of capital investment was observed in the regions of northern Italy, southern Germany, the Netherlands, Finland and Catalonia in Spain (Figure 2.2). Regions in Central and Eastern Europe generally are not very attractive to the inflow of new capital to industry. However, the more attractive regions (Czech Republic, western Slovakia, Silesia and Lower Silesia) are much closer to the EU average.

A detailed analysis of Polish provinces in this regard points to high rates of industrial investment relative to population in the voivodeships of Silesia (Śląskie), Wielkopolskie and Lower Silesia (Dolnośląskie). These rates are lower in the voivodeships of eastern Poland and in Małopolskie Voivodeship. (Figure 2.3)

Investment data show that among Polish provinces with the high rates of investment, the fastest growing are Łódzkie, Kujawsko-Pomorskie and Lubelskie voivodeships (Figure 2.4). Lower Silesia (Dolnośląskie), Wielkopolskie and Silesian (Śląskie) voivodeships are characterised by an average rate of investment growth, but the high annual average value of investment reflects the relatively rapid development of industry in these regions. Attention should also be paid to low growth in Mazovian (Mazowieckie) Province, which shows high rates of investment, but the rates indicates a decline in the relative position of this region. A relatively large number of business investments have been noted in selected provinces in Eastern Poland including Lubelskie Voivodeship, Warmian-Masurian Voivodeship, and Podlaskie Voivodeship. However, due to the low value of investment, even while maintaining its investment rate, a considerable time is needed before these regions reach a significant level of investment.

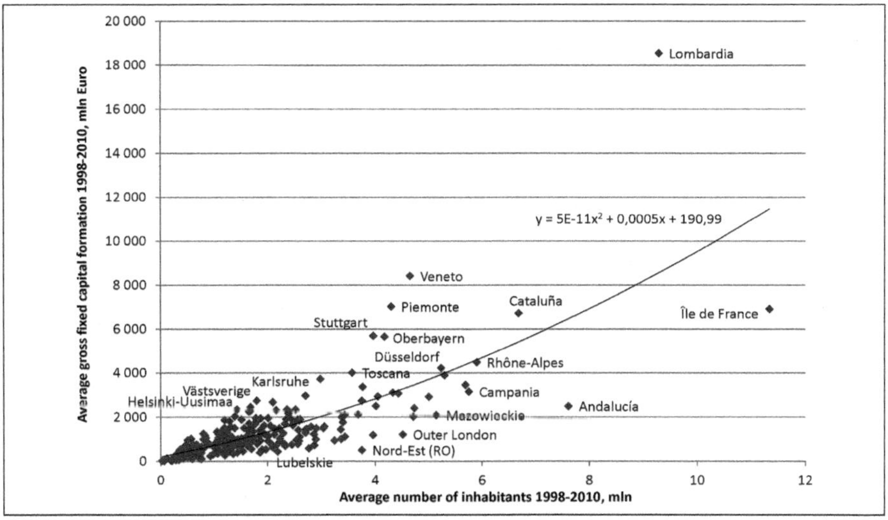

Figure 2.2 Average annual gross fixed capital formation in industry in relation to average population in EU regions (NUTS 2, 1998–2010)

Source: authors' elaboration based on Eurostat Database.

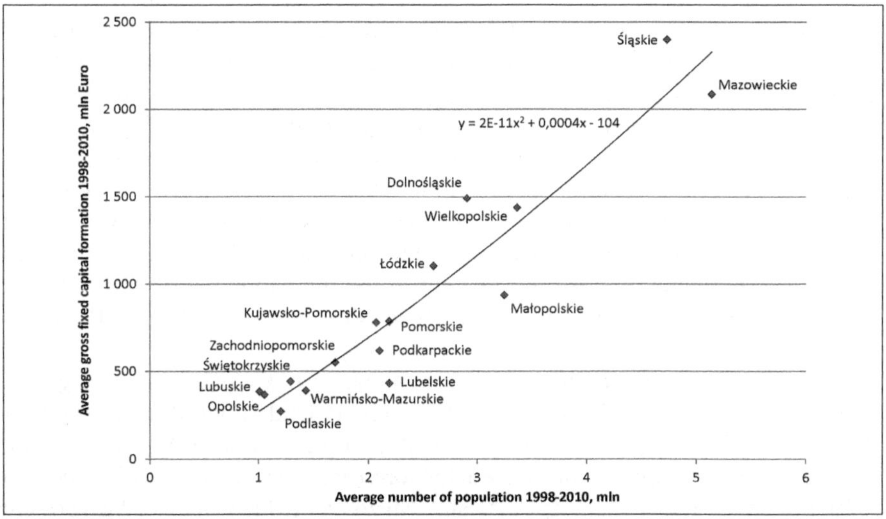

Figure 2.3 Average annual gross fixed capital formation in industry in relation to average population in Polish voivodeships (NUTS 2, 1998–2010)

Source: authors' elaboration based on Eurostat Database.

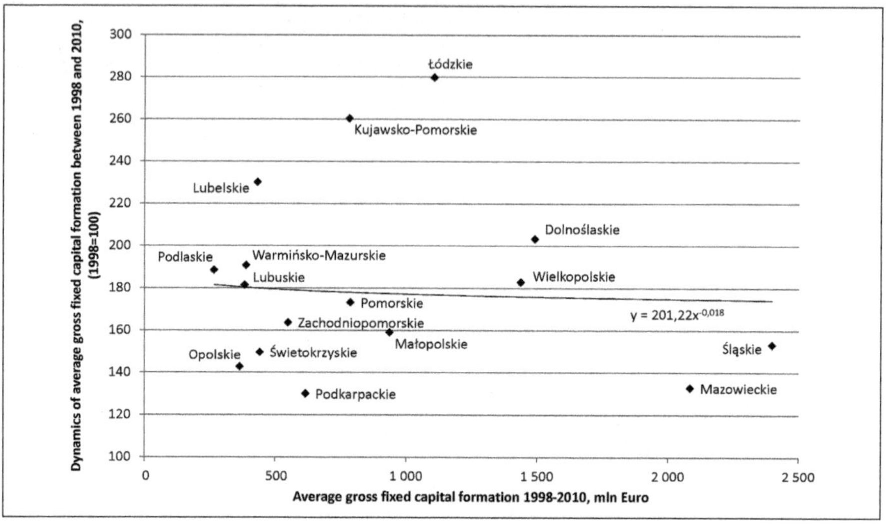

Figure 2.4 Average annual gross fixed capital investment in industry and rates of change in Polish voivodeships (NUTS 2, 1998–2010)

Source: authors' own elaboration based on Eurostat Database.

The analysis of the value of investment per capita for each year from 1998 to 2010 allows for the identification of three characteristic periods (Figure 2.5). In the period 1998–2003, before Poland's entry into the European Union, this indicator had not undergone great changes in most voivodeships – fluctuating at a stable level. A high growth of investment in industry per capita, however, is visible from 2004 to 2008, when robust growth in investment took place. This concerns mainly traditional industrialised regions such as Silesian (Śląskie), Lower Silesian (Dolnośląskie), Łódzkie and Wielkopolskie provinces. The effect of the most recent global economic crisis is apparent in the third time period from 2008 to 2010. The rate of investment falls abruptly in 2009, but rebounds in 2010 in most Polish voivodeships.

The lowest investment in industry per capita, despite a relatively high growth rate, can be found in Małopolska and Eastern Poland (Podlaskie, Warmińsko-Mazurskie, Lubelskie, Podkarpackie). In 2010 Lower Silesian (Dolnośląskie, and Silesian (Śląskie) voivodeships were at the forefront and were the most competitive of the regional labour markets in Poland. Mazovian (Mazowieckie) Voivodeship lost its leading position – it lost its leadership status in terms of investment in 1998, dropping to 5th place in 2010. If this trend is maintained, the provinces will continue to lose its position in terms of investment inflows per capita.

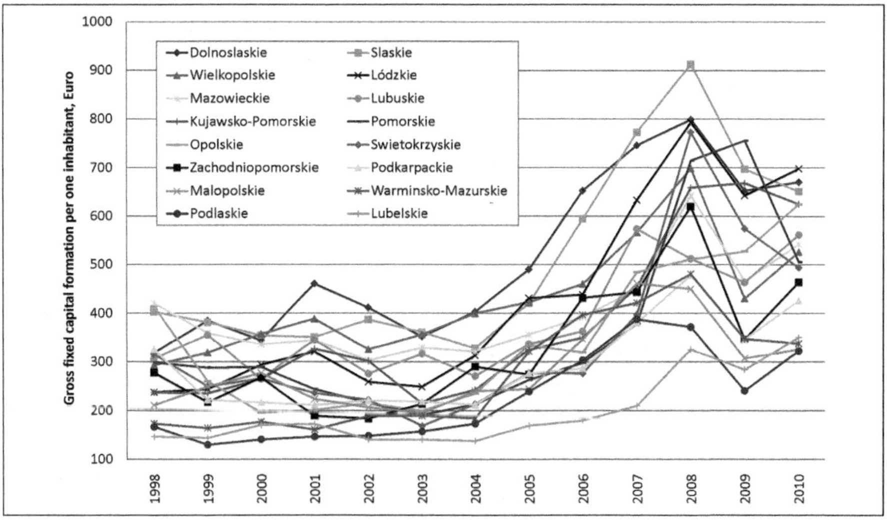

Figure 2.5 Changes in gross fixed capital investment per capita in Polish voivodeships (NUTS 2, 1998–2010)

Source: authors' own elaboration based on Eurostat Database.

The Role of R&D in the Economy

It is generally accepted that the level of investment in fixed assets in production is a sign of business innovation, which is highly influenced by the intensity of the research and development conducted by industry and its collaborating institutions. OECD studies suggest that most companies introducing product and technological innovations are also involved in research and development. An important element related to the development of industry and regional economy is the development of the R&D sector. This was noted by – among others – Malecki and Cheshire (2004) who argued that the key element in a dynamic regional economy is the presence of R&D activity.

Industrial monitoring conducted for the European Commission (The 2010 EU Industrial R&D 2010), based on investment data in research and development at the 1,000 largest companies in the EU shows that R&D investment[1] is highly concentrated

1 Research and Development (R&D) investment in The 2010 EU Industrial R&D Investment Scoreboard is the cash investment funded by the companies themselves. It excludes R&D undertaken under contract for customers such as governments or other companies. It also excludes the companies' share of any associated company or joint venture R&D investment. Being that disclosed in the annual report and accounts, it is subject to the accounting definitions of R&D.

at companies whose headquarters are located in the largest of European countries such as Germany, France and the United Kingdom[2] (Figure 2.6a).

The total value of R&D investment is more than 88bn euro, which is 68 per cent of total expenditures (almost 130bn euro) at the companies studied. R&D investment is also very high in countries such as the Netherlands, Sweden, Italy and Finland. In Central and Eastern Europe, R&D expenditures are very low due to the relative lack of large corporations. By contrast, the small country of Malta (Figure 2.6b) has the highest R&D investment rate per employee in Europe. However, only two out of the 1,000 studied corporations are based in Malta, which means that the country's high R&D investment rate may be based on very specific circumstances. High R&D investment rates per employee are also found in Finland, Denmark, Hungary, Slovenia, Greece, Germany, Sweden and the Netherlands. This underscores the importance of R&D at domestic corporations and affects the innovation level.

Earlier research has shown that the structure of R&D expenditures in Poland and other countries in Central and Eastern Europe is rather unique. Investment data indicate that close to 83 per cent of R&D spending in Poland in 2009 went to purchase fixed assets such as machines, buildings, and land. Only about 10 per cent of the spending went towards research and development (Rachwał 2012). Investment of this type usually generates higher quality products and lower costs, but does not lead to new and original products traditionally associated with research and development. This is a strategy that is typical of countries making the transition to a modern economy. Central and Eastern Europe is one region making such a transition.

Many companies in this region are either still in the process of technological restructuring or have not yet achieved a strong position in the marketplace, which makes spending on new machinery a priority in order to increase product quality. Research and development are less important to companies at this stage of business development. This is especially true at Polish companies as well as other Eastern European companies acquired by foreign investors. Large international corporations with large R&D departments that take over a given Polish company tend to eliminate R&D work at the newly acquired company. New products are then imported to Poland from the parent corporation and its R&D facilities. Rarely do international corporations shift their R&D work to Central and Eastern Europe. In short, R&D expenditures increase in the corporation's home country and not in Poland or other countries in Eastern Europe.

Not only are business expenditures on research and development small in Poland, so are government expenditures. In effect, the percentage of Poland's per

2 The focus of the Scoreboard on R&D investment as reported in group accounts means that the results can be independent of the location of the R&D activity. The Scoreboard indicates the level of R&D funded by companies, not all of which is carried out in the country in which the company is registered. It may be assumed that most R&D work occurs in countries where a company is registered and has its headquarters.

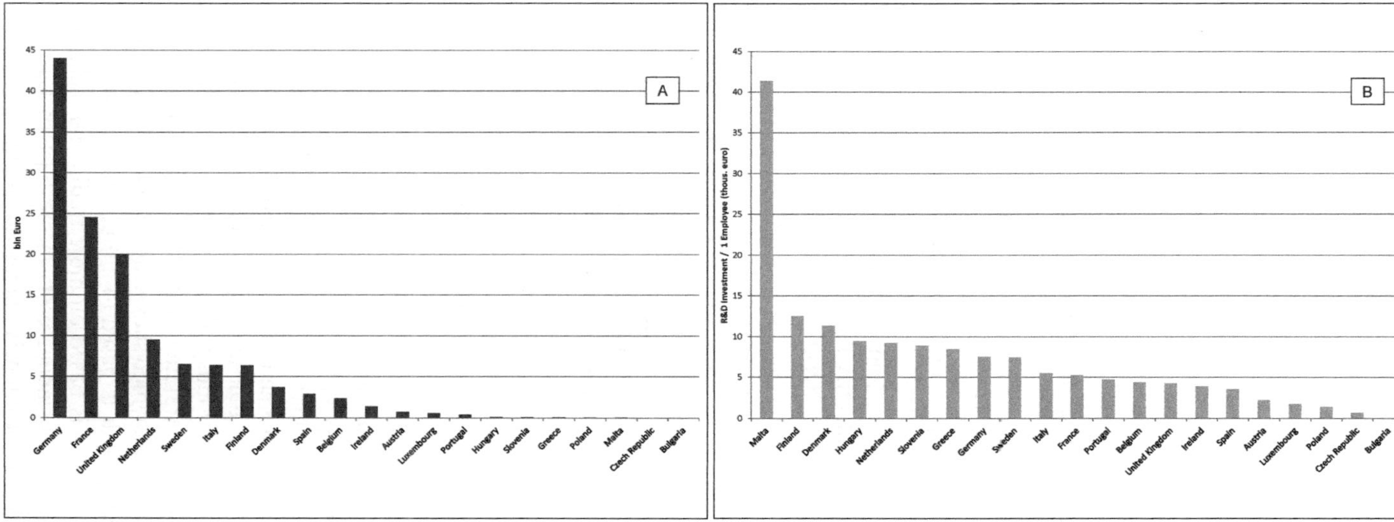

Figure 2.6 **R&D investment at 1,000 of the largest corporations in the EU listed by country (a) Investment value listed in billion EUR (b) Investment value per one employee (,000 EUR)**

Source: authors' own work based on The 2010 EU Industrial R&D Investment Scoreboard (2010).

capita GDP used for R&D is one of the smallest in the developed world. In 2008 Poland's R&D investment index was more than three times smaller than that for the European Union (approx. 2 per cent) and more than six times smaller than that for R&D leaders such as Sweden and Finland (3.7 per cent). This difference is even more pronounced when expressed on a per capita basis. The investment gap between Poland's R&D and that of the best developed nations in the world is immense – more than ten times less spending that Sweden and Luxembourg. Even the gap between Poland and other countries in Central and Eastern Europe (Hungary, Slovakia, Czech Republic) is very large (Rachwał 2012).

While 62 per cent of the largest European corporations are industrial corporations, close to 83 per cent of R&D spending takes place at industrial corporations. This indicates that R&D work is much more important for industrial corporations than service corporations. One reason for this is the relatively large potential of industrial corporations. Another reason is the ability of such corporations to stimulate innovation based largely on advanced research and development. In effect, universities and other research institutions and companies enter the picture. Even small and midsize companies join the chain of innovation. ICB (Industry Classification Benchmark) data identify three types of industrial activity.

The most significant role of R&D can be observed in the automotive industry, followed by the pharmaceuticals and biotechnology sector and the telecommunications equipment sector (Figure 2.7). Among the 1,000 surveyed companies, those that are part of the three aforementioned sectors are responsible for almost 45 per cent of total R&D expenditures. Other key sectors include: aerospace and defence, chemicals, electrical components and equipment. The six sectors account for nearly two-thirds of all R&D investment.

An analysis of the connections and relationships between the development of R&D and the development of the high-tech industry as well as GDP growth will make it possible to draw conclusions about contemporary changes taking place in European regions. From the point of view of Poland's development, it is also important to identify potential development paths for economic structures present in Polish regions.

Employment growth data for R&D sectors in 2004–2010 (Figure 2.8), analysed in this chapter, do not show a correlation (coefficient: –0.13) with data on employment in high-tech sectors in 2010. This is due to the various models of development characteristic of developing regions.

The largest employment growth in the R&D sector is characteristic of regions such as Sjaeland (DK), Iles Balears and Cantabria (ES), Norte, Centro and Lisboa (PT). This demonstrates the fact that investment here is more focused on new technologies, which in the case of industrial production, only makes up for higher production value added. From Central Europe the fastest growing regions are found in Romania, Bulgaria and Poland, which shows the low base from which they have started. Furthermore, their growth is based to a much lesser extent on R&D, as new investment in these regions is based on more traditional technologies. It is also a consequence of the lowest labour costs in the Eastern European Union, which favours the influx of labour-intensive industrial activity.

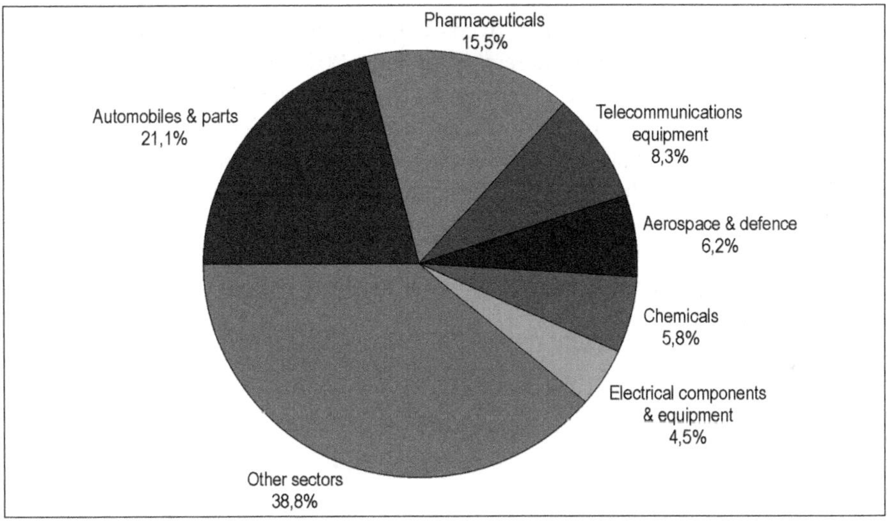

Figure 2.7 R&D investment at 1,000 of the largest corporation in the EU in 2009 – by ICB sector.

Source: authors' own elaboration based on The 2010 EU Industrial R&D Investment Scoreboard (2010).

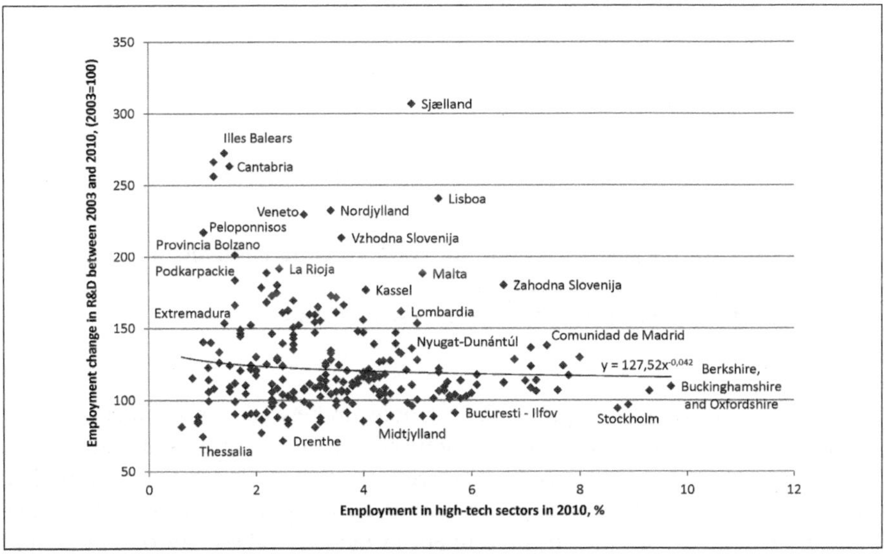

Figure 2.8 Changes in employment in R&D vs. employment in high-tech sectors in EU regions (NUTS 2, 2004–2010)

Source: authors' own elaboration based on Eurostat Database.

An analysis of the relationship between employment changes in R&D and GDP growth for the studied European regions in 2004–2010 shows that this relationship is also not significant (Figure 2.9). High GDP growth rates during this period were characteristic, above all, for the countries of the former 'Eastern Bloc'. Therefore, it can be concluded that development was based largely on investment in more traditional activities that do not necessitate a large increase in employment in the R&D sector. The highly developed areas of the 'old EU' are transforming their economic structures in the direction of new technologies, but this often does not involve a considerable increase in GDP in these regions (Auvergne, Eastern Germany, Spain).

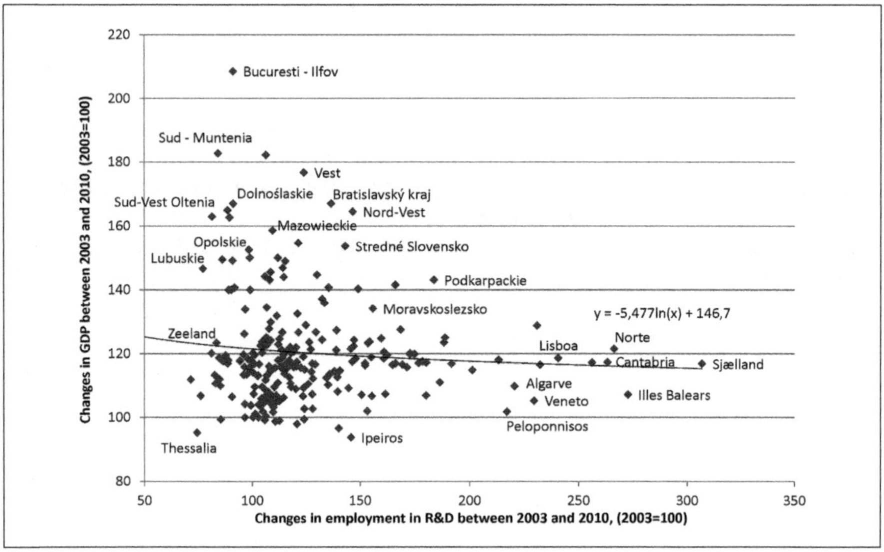

Figure 2.9 Employment changes in R&D vs. GDP changes in EU regions (NUTS 2, 2004–2010)

Source: authors' own elaboration based on Eurostat Database.

At the level of European regions, the relationship between employment in high-technology sectors and investment in industry does not show a clear pattern. Two main types of relationships can be observed. On the one hand, there are regions with a very high share of individuals employed in the R&D sector, which does not translate into high rates of investment in industry. These are often the main economic regions of countries with developed specialised services (Stockholm, Berlin, London), and regions with high-performing universities (Berkshire, Buckinghamshire, Oxfordshire, Utrecht). On the other hand, there are regions where one can observe a clear relationship between employment in the high-tech sector and investment in industry. The best examples include Upper Bavaria, Ile de France, Stuttgart and Rhône-Alpes, where the focus is the manufacturing of technologically advanced products, mainly in the automotive, aviation, electronics and pharmaceutical industries. A separate group of regions is that with a relatively large influx of industrial investment in fixed assets, but this process is not correlated with a high percentage of employees in the high-tech sector. Most often, these are old industrial districts, which introduce new and productive forms of investment. However, they are not related to the development of high technologies. Examples include Catalonia, Valencia and the Nord-Pas-de-Calais.

In the case of Polish regions, the above development models for high-tech sectors and industry will further demonstrate the two different paths (Figure 2.10). This is particularly evident for the ratio of fixed capital investment per capita in relation to employment in high-tech sectors. There is a considerable difference between highly industrialised regions including Świętokrzyskie, Lubuskie, Łódzkie, Wielkopolskie, Silesian (Śląskie) and Lower Silesian (Dolnośląskie) voivodeships, where fixed investment growth in industry will increase employment in high-tech sectors. In regions with a lower share of industry in the economy, this dependence is less clear. Mazovian (Mazowieckie), Pomeranian (Pomorskie), Małopolskie, Podlaskie and Lubelskie voivodeships have a significantly lower value of investment in fixed assets in industry compared to a relatively high share of employment in high-tech sectors. This reflects a greater role of universities and R&D institutions operating characterised by

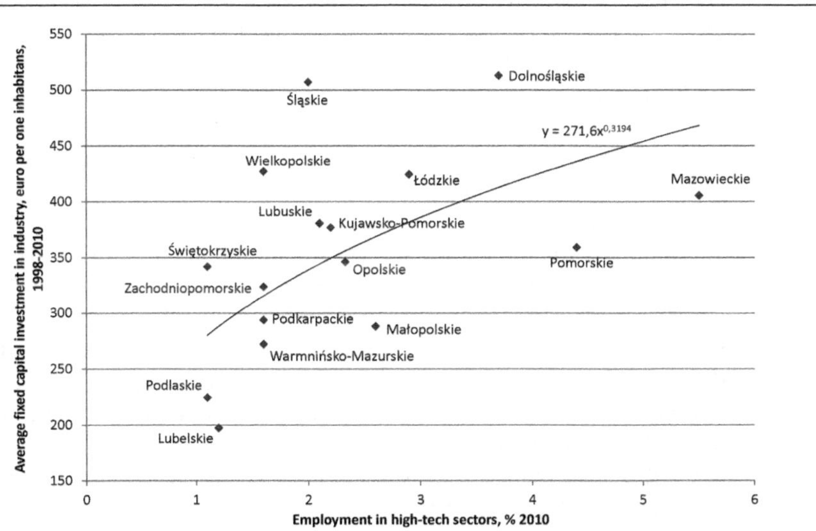

Figure 2.10 Employment in high-tech vs. fixed capital investment in industry in Polish voivodeships (NUTS 2, 1998–2010)

Source: authors' own elaboration based on Eurostat Database.

weak interactions with industry, which to some extent is the result of the vulnerability of collaboration in commercialising the results of scientific research. The issue of poor collaboration between universities and industry, characteristic of the countries of Central and Eastern Europe and the Commonwealth of Independent States, including emerging countries, was also stressed by other researchers (for example Meske 2004; Radosevic 2011).

Conclusion

Many regions currently undergoing to the transition to an innovative economy are experiencing growth based on industrial development that is usually not linked to an increase in research and development. However, not all regions should base their high-tech development exclusively on large capital investment in fixed assets in industry.

Despite the major role that industry plays in innovation and R&D work, no clear relationship was observed between industrial spending on innovation, research and development work, and economic growth measured via GDP.

Five distinct paths of R&D growth have been identified in relation to GDP growth in the EU:

1. Low R&D growth and relatively fast GDP growth, largely thanks to increasing labour-intensive production volumes. This is characteristic of regions with very low labour costs and a relatively long history of manufacturing (for example Romania).
2. Low R&D growth and little GDP growth in regions with high employment in the R&D sector. This is especially true of university research parks in Great Britain, which feature little industry. It may be inferred that such regions are operating at their R&D maximum under the current social and economic conditions.
3. Low R&D growth and little GDP growth in regions with extensive R&D facilities linked with large local industrial companies. The industrial regions of southern Germany are an example.
4. High R&D growth and relatively large GDP growth linked with the industrial sector. This is often the case in Central Europe – a region that successfully made the transition to a market economy by restructuring old industrial conglomerates.
5. High R&D growth and relatively large GDP growth largely unrelated to the industrial sector. This includes large metropolitan areas with major universities.

In Poland, the following two R&D growth paths are the most important:

1. R&D growth related to industrial investment taking place in industrial regions undergoing major sector restructuring – away from traditional manufacturing and towards technologically advanced manufacturing (for example Silesia and Lower Silesia).
2. R&D growth in regions with a growing service sector – especially metropolitan areas with major universities (Mazowieckie, Malopolskie). However, other regions following this path but lacking certain key features may experience barriers to R&D growth. Key features include large metropolitan areas and well-established major universities and research institutions.

Other areas of Poland, especially the country's northeast, feature less industry and cities without developed metropolitan areas that serve as regional economic centres. This is a barrier to research and development. This is especially true when the financial resources for R&D and higher education available from the government are very limited.

Empirical analysis of selected R&D indicators and industrial innovation indicators points to the changing role of industry in regional development in the era of the knowledge economy. This issue demands further study and new theoretical treatments. The innovative manufacturing company – regional development model described earlier in this chapter may be a useful starting point for new theoretical concepts in this area of research.

References

Amin, A. and Thrift N. 1995. *Globalization, Institutions, and Regional Development in Europe*. New York: Oxford University Press.

Asheim, B.T. 1996. Industrial districts as 'learning regions': A condition for prosperity. *European Planning Studies*, 4(4), 379–400.

Boix, R. and Trullén, J. 2010. Industrial districts, innovation and i-district effect: territory or industrial specialization? *European Planning Studies*, 18(10), 1707–29.

Czyż, T. 1998. Polaryzacja rozwoju regionalnego w okresie transformacji społeczno-gospodarczej w Polsce, in *Przemiany społeczno-gospodarcze Polski lat dziewięćdziesiątych*, edited by J. Parysek and H. Rogacki, Poznań: Bogucki Wydawnictwo Naukowe, 47–63.

Domański, B. 2000. Some aspects of the development of Polish manufacturing in the perspective of knowledge-based economy, in *The Knowledge-Based Economy. The European Challenges of the 21st Century*, edited by A. Kukliński, Warsaw: State Committee for Scientific Research, 281–7.

Domański, B. 2001. *Kapitał zagraniczny w przemyśle Polski. Prawidłowości rozmieszczenia, uwarunkowania i skutki.* Kraków: Instytut Geografii i Gospodarki Przestrzennej UJ.

Domański, B. 2003. Industrial change and foreign direct investment in the postsocialist economy: the case of Poland. *European Urban and Regional Studies*, 10(2), 99–118.
Domański, B. 2006. Polski przemysł na tle przemysłu Europy Środkowej i Wschodniej. *Prace Komisji Geografii Przemysłu Polskiego Towarzystwa Geograficznego*, 8, 27–36.
Domański, B. and Gwosdz, K. (ed.) 2005. *Dziesięć lat doświadczeń pierwszej polskiej specjalnej strefy ekonomicznej. Mielec 1995–2005*. Kraków: Instytut Geografii i Gospodarki Przestrzennej UJ, ARP o. Mielec.
Domański, B. and Gwosdz, K. 2008. Efekty mnożnikowe w rozwoju lokalnym i regionalnym, in *Region społeczno-ekonomiczny i rozwój regionalny*, edited by J.J. Parysek and T. Stryjakiewicz, Poznań: Bogucki Wydawnictwo Naukowe, 217–35.
Domański, B. and Gwosdz, K. 2010. Multiplier effects in local and regional development. *Quaestiones Geographicae*, 29(2), 27–38.
Domański, R. 1997. *Przestrzenna transformacja gospodarki*. Warszawa: Wydawnictwo Naukowe PWN.
Garnick, D.H. 1970. Differential regional multiplier models. *Journal of Regional Science*, 10, 35–47.
Gassmann, O. and von Zedtwitz, M. 1998. Organization of industrial R&D on a global scale. *R&D Management*, 28, 147–61.
Gassmann, O. and von Zedtwitz, M. 1999. New concepts and trends in international R&D organization. *Research Policy*, 28, 231–50.
Gierańczyk, W. 2003. Wybrane aspekty konkurencyjności polskiego przemysłu w dobie globalnych wyzwań rozwoju. *Prace Komisji Geografii Przemysłu Polskiego Towarzystwa Geograficznego*, 6, 77–86.
Gierańczyk, W. 2008. Handel zagraniczny wyrobami zaawansowanej technologii a społeczeństwo informacyjne Europy, in: *Problematyka XXIV Międzynarodowej Konferencji Naukowej nt. Funkcje przemysłu i usług w kształtowaniu społeczeństwa informacyjnego*, edited by Z. Zioło and M. Borowiec, Warszawa-Kraków: Komisja Geografii Przemysłu PTG, Zakład Przedsiębiorczości i Gospodarki Przestrzennej IG UP, 22–3.
Gierańczyk, W. 2009a. Innowacyjność jako główny filar społeczeństwa informacyjnego Europy. *Prace Komisji Geografii Przemysłu Polskiego Towarzystwa Geograficznego*, 13, 82–94.
Gierańczyk W. 2009b. Przemysł zaawansowanej technologii a konkurencyjność przemysłowa krajów UE, in *Problematyka XXV Międzynarodowej Konferencji Naukowej nt. Procesy transformacji przemysłu i usług w regionalnych i krajowych układach przestrzennych,* edited by Z. Zioło and M. Borowiec, Warszawa-Kraków: Komisja Geografii Przemysłu PTG, Zakład Przedsiębiorczości i Gospodarki Przestrzennej IG UP, 36–7.
Gierańczyk, W. 2010. Development of high technologies as an indicator of modern industry in the EU. *Bulletin of Geography, Socio-economic Series*, 14, 23–36.

Gierańczyk, W. and Rachwał T. 2012. Structural changes in the industry of Poland against the background of eastern European Union states. *Quaestiones Geographicae*, 31(2), 83–93.

Gorzelak, G. 1993. Regionalizm i regionalizacja w Polsce na tle europejskim, in *Czy Polska będzie państwem regionalnym?* edited by G. Gorzelak, B. Jałowiecki. Studia Lokalne i Regionalne, Warszawa: Europejski Instytut Rozwoju Regionalnego i Lokalnego Uniwersytetu Warszawskiego, 9, 45–70.

Gorzelak G., 2000, Trwałość i zmiana: historia, transformacja i przyszłość polskich regionów. *Ekonomista*, 6, 1–18.

Grabher, G. 1994. The disembedded regional economy: the transformation of East German industrial complexes into western enclaves, in *Globalization, Institutions, and Regional Development in Europe*, edited by A. Amin and N. Thrift. Oxford: Oxford University Press, 177–95.

Gwosdz, K. 2011. Efekty mnożnikowe funkcjonowania firm na terenie gminy Niepołomice, in *Sukces polityki proinwestycyjnej. Niepołomice 1990–2010. Wnioski dla władz lokalnych*, edited by W. Jarczewski and M. Huculak, Kraków: Instytut Rozwoju Miast, 129–40.

Hansen, T. and Winther, L. 2011. Innovation, regional development and relations between high- and low-tech industries. *European Urban and Regional Studies*, 18, 321–39.

Kukliński, A., Mync, A. and Szul, R. 1997. *Polska przestrzeń na przełomie XX i XXI wieku*. Warszawa: Uniwersytet Warszawski, Europejski Instytut Rozwoju Regionalnego i Lokalnego.

Malecki, E.J. and Cheshire, P.C. 2004. Growth, development, and innovation: A look backward and forward. *Papers in Regional Science*, 83, 249–67.

Mattes, J. 2012. The regional embeddedness of multinational companies. A critical perspective. *European Planning Studies*, 21(4), 433–51.

Meske, W. (ed.) 2004, *From System Transformation to European Integration. Science and Technology in Central and Eastern Europe at the Beginning of the 21st century*. Munster: LIT Verlag.

Micek, G. 2011. Estimating multiplier effects on the local scale. *Acta Universitatis Lodziensis. Folia Oeconomica*, 252, 175–90.

Parysek, J.J. 1998. Efekty procesu transformacji społeczno-ekonomicznej w Polsce. Bilans okresu 1989–1996, in *Przemiany społeczno-gospodarcze Polski lat dziewięćdziesiątych*, edited by J.J. Parysek and H. Rogacki, Poznań: Wydawnictwo Naukowe Bogucki, 25–45

Pavlinek, P. and Smith, A. 1998, Internationalization and embeddedness in East-Central European transition: the contrasting geographies of inward investment in the Czech and Slovak Republics. *Regional Studies*, 32(7), 619–38.

Porter, M.E. 1990, *The Competitive Advantage of Nations*. New York: Free Press.

Porter, M.E. 1998, *On Competition*. Boston: Harvard Business School Press.

Porter, M.E. 2003, The economic performance of regions. *Regional Studies*, 37(6–7), 549–78.

Porter, M.E. 2008, *On Competition*, updated and expanded edition. Boston: Harvard Business School Publishing.

Rachwał, T. 2009. Changes of industry in the countries of Central and Eastern Europe under conditions of economic transformation and European integration, in *Countries of Central & Eastern Europe Versus Global Economic Crisis*, edited by J. Kitowski, Geopolitical Studies, 15, Warszawa: Instytut Geografii i Przestrzennego Zagospodarowania PAN, 133–64.

Rachwał, T. 2010. Struktura przestrzenna i działowa przemysłu Polski na tle Unii Europejskiej w dwudziestolecie rozpoczęcia procesów transformacji systemowej. *Prace Komisji Geografii Przemysłu Polskiego Towarzystwa Geograficznego*, 16, 105–24.

Rachwał, T. 2011. Transformations of the employment structure as an expression of the transformation of the Polish industry against the background of the European Union. *Bulletin of Geography, Socio-economic Series*, 15, 5–25.

Rachwał, T. 2012. Innowacyjność przedsiębiorstw przemysłowych jako czynnik rozwoju miast, in *Miasto innowacyjne. Wiedza – Przedsiębiorczość – Marketing*, edited by Z. Makieła and A. Szromnik, Studia KPZK PAN, CXLI, Warszawa: KPZK PAN, 135–52.

Rachwał, T. and Boguś, M. 2012a. Konkurencyjność przedsiębiorstw przemysłowych w Polsce w ujęciu regionalnym na tle innych krajów Unii Europejskiej, in *Przekształcenia struktur regionalnych*, edited by P. Raźniak, Kraków: Wyd. Krakowskiej Akademii A. Frycza-Modrzewskiego, 9–38.

Rachwał, T. and Boguś, M. 2012b. The Competitiveness of industrial enterprises in European Union countries, in *Current problems of Competitiveness Improvement in National Economies and Enterprises*, edited by B. Krstić and A. Chodyński, Niš: Faculty of Economics – University of Niš, 133–56.

Rachwał, T. and Wiedermann, K. 2008. Multiplier effects in regional development: The case of the motor vehicle industry in Silesian voivodeship (Poland). *Quaestiones Geographicae*, 27B/1, 67–80.

Rachwał, T., Wiedermann, K. and Kilar, W. 2008a, Changes in differentiation of Polish regions' industrial potential within the European Union, in *New Functions of Rural and Industrial Space in Central and Eastern Europe*, edited by K. Czapiewski and T. Komornicki, Europa XXI, 17, Warsaw: Institute of Geography and Spatial Organization – Polish Academy of Sciences, Academic Division – Polish Geographical Society, 145–58

Rachwał, T., Wiedermann, K. and Kilar, W. 2008b. Wydajność i koszty pracy jako czynniki konkurencyjności przemysłu regionów Polski w Unii Europejskiej w ujęciu regionalnym, in *Przekształcenia regionalnych struktur funkcjonalno-przestrzennych 'Europa bez granic – nowe wyzwania'*, edited by D. Ilnicki and K. Janc, Rozprawy Naukowe Instytutu Geografii i Rozwoju Regionalnego Uniwersytetu Wrocławskiego, 3, Wrocław: Instytut Geografii i Rozwoju Regionalnego UW, 79–89.

Rachwał, T., Wiedermann, K. and Kilar, W. 2009. Rola przemysłu w gospodarce układów regionalnych Unii Europejskiej. *Prace Komisji Geografii Przemysłu Polskiego Towarzystwa Geograficznego*, 14, 31–42.

Radosevic, S. 2011. Science–industry links in Central and Eastern Europe and the Commonwealth of Independent States: conventional policy wisdom facing reality. *Science and Public Policy*, 38(5), 365–78

Storper M., 1997, *The Regional World: Territorial Development in a Global Economy*. New York: Guilford Press.

Stryjakiewicz, T. 1999. *Adaptacja przestrzenna przemysłu w Polsce w warunkach transformacji*. Poznań: Wydawnictwo Naukowe UAM.

Szlachta, J. 1993. *Poland's Regional Development under Economic Transformation*, Warsaw: Friedrich Ebert Stiftung.

The 2010 EU Industrial R&D Investment Scoreboard. 2010. Luxembourg: European Commission.

Wiedermann, K. 2007. Regionalne efekty mnożnikowe rozwoju przemysłu motoryzacyjnego w aktywizacji gospodarczej województwa śląskiego. *Przedsiębiorczość – Edukacja*, 3, 24–34.

Wiedermann, K. 2008. Koncepcja efektów mnożnikowych w wyznaczaniu wpływu przedsiębiorstw na otoczenie społeczno-gospodarcze. *Prace Komisji Geografii Przemysłu Polskiego Towarzystwa Geograficznego*, 11, 98–106.

Zioło Z. 2009, Model badań procesu transformacji elementów w przestrzeni geograficznej, in *Współczesne problemy przemian strukturalnych przestrzeni geograficznej,* edited by I. Jażewicz, Słupsk: Akademia Pomorska, 103–17.

Zioło, Z. 1996. Model funkcjonowania przestrzeni geograficznej i jego znaczenie dla gospodarki przestrzennej, in *Gospodarka, przestrzeń,* środowisko, edited by U. Wich, Lublin: Wyd. UMCS, 239–50.

Zioło, Z. 1999. Model funkcjonowania przestrzeni geograficznej jako próba integracji badań geograficznych, in *Geografia na przełomie wieków – jedność w różnorodności*, edited by A. Lisowski, Warszawa: Wydział Geografii i Studiów Regionalnych Uniwersytetu Warszawskiego, 122–31.

Zioło, Z. 2003. Kształtowanie się przedsiębiorstw przemysłowych w procesie globalizacji. *Prace Komisji Geografii Przemysłu Polskiego Towarzystwa Geograficznego*, 6, 9–25.

Zioło, Z. 2006. Zróżnicowanie światowej przestrzeni przemysłowej w świetle koncentracji siedzib zarządów wiodących korporacji. *Prace Komisji Geografii Przemysłu Polskiego Towarzystwa Geograficznego*, 8, 9–26.

Chapter 3
Shortcomings and Weaknesses in Understanding and Measuring Knowledge Interactions

Grzegorz Micek

Knowledge interactions[1] have become one of the core topics of contemporary economic debate. This is not only the internal knowledge and competences that constitute companies' potential to innovate. Dating back to Alfred Marshall's contribution, knowledge flows from external sources have been treated as a key mechanism for agglomeration and consequently innovation. Expósito-Langa, Molina-Morales and Capó-Vicedo (2011) argue that the greater the absorption of knowledge flows, the greater the innovation capacity for the company. However, there is the consensus that knowledge interactions substantially differ by sector, knowledge base, regional innovation systems characteristics (Tödtling, Lengauer and Höglinger 2011) and also by knowledge dissemination channels. Asheim, Smith and Oughton (2011) raise the crucial question of how knowledge is transferred between actors. Even if it is not very difficult to identify channels of knowledge interactions, it is certainly a more challenging task to measure the magnitude of knowledge flows in different channels. Krugman (1991) argues that it is impossible to measure knowledge interactions, because they leave no paper trail by which they can be measured or tracked. On the other hand, an opposite strand of research points out that knowledge flows may be traceable and measurable via new product introductions (Jaffe and Henderson 1993, Feldman 2003).

Paradoxically, research quality is not enhanced by increasing the number of papers on knowledge-related issues. This results in a multiplicity of knowledge-related terms (often treated as synonyms) and untranslatability of some variables into real knowledge flows. This chapter critically refers to findings in terms of definitions and measurement of knowledge interactions through a review of the most recent papers in the field of broadly understood economic geography (including regional and urban studies). It contributes to the existing body of research by identifying overuses of knowledge interactions measurement. The next section identifies knowledge interaction channels and ends with a critical review of the

1 The term 'knowledge interactions' is used in the chapter to describe six different terms: 'knowledge flows', (non–pecuniary) 'spillovers', 'exchange', 'transfers', 'diffusion', 'dissemination' and 'sourcing'.

term used to denote knowledge interactions. Research methods used in the chapter are briefly discussed in the section that follows. This is then followed by a review of the most recent research literature focused on the terms and methods used. Finally, a discussion of the shortcomings of research on knowledge interactions is presented and key weaknesses of such research are identified.

Knowledge Interaction Channels

Knowledge is shared in different forms and modes. Knowledge transmission channels have been comprehensively captured by Martin and Moodysson (2011), who divided them into three groups: monitoring, mobility, collaboration. Monitoring of scientific papers or reports is one of the examples of a search for knowledge outside the firm, but in contrast to collaboration, the interaction with external sources is indirect. Mobility articulates in retrieving knowledge resources through spin-off decisions or recruitment of key employees from other organizations including universities, technical colleges, and firms in the same or related industries (Martin and Moodysson 2011, 2013). Collaboration is based on direct interaction and might take various forms and magnitude from long-lasting formal contacts as part of research or business projects to informal, often temporary interactions. These various forms of collaboration as well as mobility of individuals shape knowledge networks (Martin and Moodysson 2013). Boschma and Frenken (2011: 301) argue 'the social network ties between moving individuals and their former colleagues or between previous collaborators may well remain a channel for knowledge exchange'.

More specifically, knowledge dissemination channels may be classified based on Martin's and Moodysson's (2011) contribution as follows:

1. Monitoring
   - Monitoring of scientific journals or reports (Martin and Moodysson 2011, Tödtling, Lengauer and Höglinger 2011, Toedtling et al. 2013)
   - Monitoring through market surveys and questionnaires (Martin and Moodysson 2011, Tödtling, Lengauer and Höglinger 2011, Toedtling et al. 2013)
2. Mobility
   - Labour mobility (Bienkowska, Lundmark and Malmberg 2011; Eriksson 2011, Tabata 2013), using knowledge acquired in former workplaces
   - Key personnel recruitment: managers (inter-locking directorates – Burris 2005) or top engineers
   - Spin-off companies (Buenstorf and Geissler 2010) set up by academics or former employees of other companies

3. Collaboration
   - Arm's length relations (pecuniary spillovers) – contract research, licensing, consulting
   - Patent collaboration: citations (Agrawal, Cockburn and McHale 2006, Alazzawi 2011), cooperation or sometimes even co-location (see for instance Ó Huallachain and Lee 2011),
   - Collaboration within industrial consortia, mutual agreements, and so on
   - Collaboration on academic and technical papers: citations or cooperation
   - Collaboration within research schemes, especially under EU Framework Programmes (Scherngell and Barber 2011, Basile, Capello and Caragliu 2012)
   - Interactions during fairs, conventions, conferences (Bathelt and Schuldt, 2008, Schuldt and Bathelt 2011, Rychen and Zimmermann 2008) designated 'temporary proximity' that enables the emergence of temporary clusters
   - Informal contacts between different companies' employees (often former classmates; during informal events, parties, and so on).

The list above is an expanded version of knowledge transfer channels used in empirical research by Tödtling, Lengauer and Höglinger (2011): contract research; licences, machinery, software; R&D collaboration; informal contacts; employment of specialists; monitoring of competitors; conferences and conventions; academic journals, magazines.

Martin and Moodysson's (2011) classification is consistent with the typology by Desrochers and Leppala (2011: 852) who identified three sets of circumstances 'through which individuals found new uses or applications for existing products and created new combinations of existing products, processes and materials: (i) by adding to, switching or adapting specific know-how to other lines of work; (ii) by observing something in another line of work and incorporating it into one's own line of work; and (iii) through formal and informal collaboration between individuals possessing different skills and working towards the creation of new products and processes'.

Tödtling, Lengauer and Höglinger (2011) present a different typology by dividing external knowledge links into categories that include market relationships such as contract research, licensing, and consulting. Another category (externalities and spillovers) entails recruitment of specialists, monitoring of competitors, participation in fairs and conferences, and reading of the scientific literature and patent specifications. In addition to the above-mentioned static relationships, Tödtling, Lengauer and Höglinger (2011) distinguish dynamic linkages based on collective learning that lead to the emergence of formal networks or informal networks.

It must be emphasized that the above-mentioned knowledge transfer channels differ in terms of grasping real knowledge flows. This especially applies to citations, which in fact indirectly and weakly reflect knowledge interactions.

Overview of Terms Used to Denote Knowledge Interactions

'Knowledge flows', (non-pecuniary) 'spillovers', 'exchange', 'transfers', 'diffusion', 'dissemination' and 'sourcing' are often used interchangeably, but they slightly differ in meaning and consequently they should differ in their usage. In this section, the understanding of these terms is discussed and contrasted.

Knowledge flows are often used to describe knowledge interactions. Unlike water, knowledge is not usually a continuous flow, and Feldman (2003: 386) argues 'is affected by discoveries and breakthroughs that provide different opportunities to realize technological advance and innovation'. The term 'knowledge flows' is used often in cases where qualitative, non-econometric analysis is being applied. However, Boschma and Frenken (2011) discuss labour mobility and treat it as one of the important channels of knowledge flows. They distinguish *knowledge flows* from *knowledge spillovers* as sources of Marshall's externalities[2] based on regional specialization (Boschma and Frenken 2011). Brown and Rigby (2010, cited in Boschma and Frenken (2011)) identify three different sources of MAR (Marshall–Arrow–Romer) externalities: knowledge spillovers, labour market pooling, and buyer–supplier relationships. In this respect, they substantially distinguish knowledge spillovers from labour flows. The terms 'knowledge externalities' and 'spillovers' are used here interchangeably (Gallie 2009, Fischer 2011).

Yet, knowledge spillovers are still treated by many researchers as a 'black box' (Boschma and Frenken 2011). Knowledge spillovers are usually defined as non-pecuniary flows that are not linked to knowledge embodied in traded capital or intermediate goods (Fischer 2011). Referring to Almeida and Kogut (1999), Gallie (2009: 40) argues that knowledge spillovers denote 'the benefit of knowledge to people, or to firms, not responsible for the original investment in the creation of this knowledge'. There exists a simplistic assumption that knowledge 'spillovers increase the level of technology for all firms in the region'. (Fischer 2011: 422). The term is more often used in empirical modelling (especially within the knowledge production function) and spatial economics than in knowledge flows.

The next term used to discuss knowledge interactions is *knowledge transfer*. It is sometimes used together with the noun 'exchange' forming the more complex term *knowledge transfer and exchange (KTE)* (Mitton et al. 2007, Campbell et al. 2011). The term 'transfer' seems to be better than 'flows', but it also implies using an institutional framework. It is frequently used to discuss industry–university relations (Huggins and Kitagawa 2013). While it is fairly obvious that data, general information (Raymond and Robinson 2013), technology (Lin et al. 2011), skills (Alazzawi 2011), competences, and capabilities (Buenstorf and Geissler 2010) are transferred between companies located within a given geographic region, it is debatable whether knowledge understood as a complex set of related information

2 Known in the literature as MAR externalities, referring to Marshall–Arrow–Romer's contribution.

is actually transferred. The same applies to *knowledge exchange* as it is obvious that information (Kemeny 2011) or ideas (Heebels 2013) are exchanged rather than knowledge itself. However, some researchers (Mulrennan, Mark and Scott 2012) argue that reciprocal transfer of knowledge, skills, and capacities understood as co-learning may occur between different stakeholders. Hence, bilateral exchange of knowledge (Martin and Moodysson 2013) may occur, but it does not entail all types of knowledge interactions, as it excludes cases where knowledge is transmitted in one direction only.

A growing number of research studies on *knowledge sourcing* are available (Sotarauta et al. 2011, Huber 2012, Tödtling et al. 2013), especially in the context of regional innovation systems (Sotarauta et al. 2011, Tödtling et al. 2013). Another use of the term is presented by Sotarauta et al. (2011) who argue that knowledge is sourced in the process of the formal education of employees. Hence, knowledge sourcing is said to be linked to the educational background of employees.

There are other terms that describe knowledge interactions. The term *knowledge circulation* has not been used in geography in recent years, although it appropriately reflects the processes described by Alfred Marshall in the atmosphere that facilitates face to face contacts in the specific milieu ('it's something in the air'). Knowledge circulation would correctly describe knowledge interactions that occur during trade fairs (Bathelt and Schuldt 2008).

Knowledge diffusion is the closest in meaning to knowledge spillovers. However, according to the hierarchical model of spatial diffusion, knowledge first spills over primarily to the largest urban centres (companies) and later to smaller sites (firms). In the knowledge spillover model, it is not stated that information cannot be shared with the spatially closest neighbours.

Knowledge dissemination is also rarely used in the ongoing debate in economic geography. Hennemann (2011) argues that transnational corporations play a crucial role in innovation and the dissemination of new ideas. The meaning of the term *knowledge assimilation* is slightly different than the above-discussed, as it depends on the reader's capacity to understand obtained information. Expósito-Langa, Molina-Morales and Capó-Vicedo (2011) introduce the term *assimilation capacity* arguing that it is a critical element in the innovation process in industrial areas.

As it may be inferred from the above-presented short review, there are multiple terms that describe knowledge interactions, which have similar or identical meaning. Similar meaning applies to knowledge dissemination, diffusion, and spillovers, even if flows should be used instead while discussing labour mobility. Knowledge exchange slightly differs from the other terms, because it relies on the sharing of knowledge between two actors.

Methods

The systematic review of the use of six different variations of knowledge interactions (spillovers, flows, diffusion, exchange, transfer, sourcing) was conducted using research journals indexed in the Social Sciences Citation Index (Web of Knowledge). The aim was to investigate the most recent articles that were published in the field of broadly understood economic geography (including regional and urban studies). Hence, papers[3] published in 'geography-' and 'urban studies-classified'[4] journals between January of 2011 and April of 2013 were selected and reviewed thoroughly in three different fields:

- terms used to discuss knowledge interactions
- methods applied
- indicators of knowledge interactions.

First, the frequency of use of the six terms related to knowledge interactions was analysed. Second, the use of qualitative and/or quantitative methods to discuss the analysed terms was investigated. Third, methods of the measurement of knowledge interactions were investigated.

Knowledge-related Terms – Selected Statistics

Initially, a total of 101 papers were selected, mainly from publications such as Regional Studies, Journal of Economic Geography, Papers of Regional Science, and European Planning Studies. Twelve of the 101 papers were rejected due to the accidental occurrence of terms related to knowledge interactions – in such cases, knowledge flows were usually mentioned only in the reference section. Ultimately, only 89 papers were used in the quantitative analysis presented below.

The first step of the literature review of the number of research papers within particular disciplines of science has given following results. Within economic geography, knowledge spillovers are most commonly (60 papers based on a query on the topic in Web of Science) used to describe knowledge interactions, especially in the framework of spatial economics and new economic geography. Geography and urban studies are ranked third in terms of the number of papers (Table 3.1) and are exceeded by economics and management studies. In terms of the number of papers, 'knowledge exchange' comes in second (n=21) and is followed by 'knowledge flows' (n=15) and 'knowledge transfer' (n=14). The latter term is used more commonly in disciplines other than geography. The term knowledge flows is popular in papers written by American researchers and is usually used in qualitative papers. Knowledge diffusion and sourcing are rarely used in the recent literature in economic geography. However, knowledge sourcing is quite a new term, and it seems that it is becoming more popular.

---

3 Proceedings papers were excluded from the analysis.

4 Most economic geography papers are published in journals classified in these two Web of Science categories.

Table 3.1 Use of six analysed topics in Geography and Urban Studies and other Web of Science disciplines

Term	Leading Web of Science discipline (number of papers since 1945)	Rank of Geography and Urban Studies in comparison with other Web of Science disciplines	Number of papers–all disciplines	Geography & Urban Studies	
				Total	*Analysed period (January 2011 to April 2013)*
Knowledge spillovers	Economics (499)	3rd place	768	183	49
Knowledge exchange	Management (94)	3rd place	406	49	21
Knowledge flows	Management (221)	5th place	438	60	15
Knowledge transfer	Management (884)	11th place	1970	65	14
Knowledge diffusion	Economics (62)	5th place	219	24	8
Knowledge sourcing	Management (48)	3rd place	73	13	8
Total			3874	394	115[a]

Note: [a] The total number in the table (n=115) does not equal the total number of papers (n=89). In some papers, several knowledge-related terms may have been used together.
Source: own compilation based on Web of Science.

The second step of the analysis was the following. For each paper, the number of occurrences of the six analysed terms was calculated (Table 3.2), and the significance of the term for the paper's line of argument was estimated. Based on these two indicators, a hierarchy of terms in each of the 89 papers was identified. Looking not only at the topic searched in Web of Science, but at the specific content of research papers, selected statistics on the number of occurrences of the six analysed terms were produced. On average, one paper included 2.7 knowledge-related terms. This confirms that the vast majority of papers contain more than one of the analysed terms. One term was used in only 20 papers. In many cases, terms are used interchangeably. This is especially true of *knowledge exchange* and *transfer*, while both terms do differ semantically. The two most popular pairs of terms in the papers analysed were *spillover-transfer* and *spillover-exchange*. *Transfer* was especially popular as an auxiliary term.

Table 3.2 Occurrence of selected six terms in analysed research papers

Rank	Spillover	Exchange	Transfer	Flow	Diffusion	Sourcing	Total[a]
1st	40	20	16	10	6	6	98
Including stand-alone terms	*9*	*5*	*3*	*1*	*1*	*1*	*20*
2nd	7	15	14	13	10	3	62
3rd	8	11	15	10	3	1	48
4th	1	6	7	4	2	0	20
5th	3	2	0	0	3	0	7
Total	**59**	**54**	**52**	**37**	**24**	**10**	**236[b]**

Notes: [a] in some papers, the number of occurrences of two or three terms was identical, so these terms were assigned the same rank; [b] number of terms used in analysed 89 papers.
Source: own compilation.

A total of 89 papers were used to conclude that three groups of knowledge interactions are described by a variety of terms. While discussing the monitoring of other organizations, researchers prefer to use the terms knowledge flows and knowledge diffusion, as opposed to knowledge spillovers. The terms *knowledge spillovers*, *knowledge transfer*, and *knowledge exchange* are often used to analyse collaboration (especially within R&D-related activities), which is the most commonly studied type of knowledge interactions. In summary, mobility seems to imply *knowledge flows*, *transfer*, and *spillovers*.

Collaboration is the most commonly studied type of knowledge interactions. Labour mobility is a much less common framework for knowledge interactions. Monitoring is hardly ever studied in the field of economic geography, because it is difficult to capture except for a few pieces of research (Tödtling, Lengauer and Höglinger 2011) that raise direct questions during interviews.

Methods Used to Discuss Knowledge-related Processes

The most common (n=31 papers) method of analysis of knowledge interactions is empirical modelling (Figure 3.1). The next most common method is papers (n=17) that show results based on surveys and/or interviews. However, such research is much less common than quantitative methods. Mixed-method research studies have become popular in contemporary economics and sociology during the last decade. This is not strongly reflected in studies on knowledge interactions that follow two distinct schools of thought: (1) dominant quantitative approach, (2) qualitative approach. It may be argued that using one approach may lead to biased results.

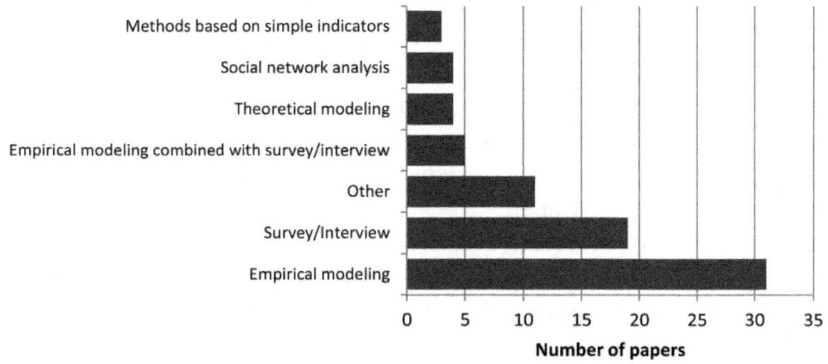

Figure 3.1 Groups of methods used to analyse knowledge interactions
Source: own compilation.

Very often the analysis of quantitative and qualitative data is done separately by different researchers. However, it appears reasonable that the results of the analysis of quantitative and qualitative data should be integrated. A good way to do it would be to link the results of social network analysis with qualitative results.

Measurement of the Magnitude of Knowledge Interactions

How exactly are knowledge flows measured? Based on the review of research papers, it must be argued that the way knowledge flows are captured varies substantially and must depend on investigated channels. The most common method of measurement of knowledge interactions is based on R&D-related variables (n=11 papers). The simplest method of identification of knowledge spillovers is based on the presence of an R&D department in a region (Alkay and Hewings 2012). However, such a presence (measured by a dichotomous variable) does not automatically mean that knowledge spillovers occur. Fritsch and Slavtchev (2011) use the number of R&D employees as a proxy for knowledge spillovers within the private sector. They argue that 'the greater the number of R&D employees, the greater the opportunity to find a suitable partner for cooperation and knowledge exchange'. A more advanced indicator of knowledge spillovers within the private sector is the share of R&D employment in that sector (Fritsch and Slavtchev 2011). The most sophisticated R&D-based variable has been used by Abel and Deitz (2012) who studied the intensity of academic research activity at colleges and universities in metropolitan areas. This variable was calculated as academic R&D expenditures per enrolled student in a metropolitan area. Abel and Deitz (2012: 675) argue that this indicator 'serves as a proxy for the research orientation of its colleges and universities, capturing the potential for knowledge spillovers'. Other R&D-related dimensions widely studied in recent years include the co-location of patents (Ó Huallachain and Lee 2011) and the presence of multiple inventors, with at least one located in a different region or country (Agostino, Laursen and Santangelo 2013). Usai (2011: 723) emphasizes the role of patents, arguing that an 'invention generated in one region may spill over and help knowledge formation in other regions'. However, the linkage between patent citations and knowledge interactions could be questioned.

The parameters of *knowledge networks* (size, density, number of ties) were used to measure knowledge interactions in eight research papers. For instance, Wenting, Atzema and Frenken (2011) treat the size of the social and professional network of an entrepreneur as a proxy for knowledge spillovers. Weiss et al. (2012) asked survey respondents to identify which entities they or their organization shares information with. The number of technology, knowledge, and market linkages (for example, collaborative science–industry agreements) was analysed by Sotarauta et al. (2011).

The third most common (n=7) method of identification of knowledge flows is based on survey questions on the importance of various external sources of business, scientific, and technological knowledge. Wang and Lin (2013) evaluated the importance of knowledge exchange in technology development with domestic and foreign firms on the basis of firms' response to a questionnaire. Huber (2012: 114) asked interviewees the following question: 'To what extent is it beneficial for your work in your current firm to have many innovative firms/research institutions located in the Cambridge region?' Tödtling, Lengauer and Höglinger (2011) classify the significance of knowledge interactions on a scale from one (very low) to five (very high) in terms of the number of contacts, the importance of knowledge transfer channels, and the similarities of exchanged knowledge. In a similar way, v Martin and Moodysson (2011, 2013) studied the perceived importance of external knowledge sources in the field of monitoring, mobility, and collaboration.

Labour mobility (pooling) is ranked third (n=7) in terms of the number of occurrences in the analysed papers. These studies include the measurement of the scale and scope of the mobility of top scientists (Schiller and Revilla Diez 2013). Within this field of studies, labour mobility is usually not treated as a direct indicator of knowledge interactions, but rather as a proxy for knowledge spillovers (Bienkowska, Lundmark and Malmberg 2011).

There are many other proxies for knowledge interactions that should be treated with caution. Jordaan and Rodriguez-Oreggia (2012) treat FDI inflows as knowledge spillovers, arguing that foreign companies may provide new technical or managerial knowledge. This line of argument seems to be an oversimplification, as there is evidence that at least some foreign companies provide only limited innovation and have been nicknamed *cathedrals in the desert* (Morris 1992, Grabher 1994, Hardy 1998). Although this 'branch plant syndrome' has been recently questioned in papers written in Western Europe (Sonn and Lee 2012), there is some evidence that the local impact of foreign companies operating in

Central and Eastern Europe on knowledge interactions is limited in some sectors (Micek, Działek and Górecki 2011).

The next proxy of knowledge interactions has been suggested by Croce and Ghignoni (2013), who argue that the percentage of university graduates at the local level can be regarded as a source of knowledge spillovers. The underlying assumption is that more educated people are more often involved in knowledge interactions.

One of the most appropriate quantitative variables of knowledge interactions was used by Potter and Watts (2011) who asked interviewed firms about knowledge their manufacturing plant valued[5] at a minimum of £1,000, or the equivalent of five days of inter-firm knowledge transfer. This enabled researchers to distinguish and assess the value of tacit knowledge acquired by companies. However, one may imagine it is difficult to obtain such data.

Discussion of the Shortcomings of Knowledge Interactions Research

There are at least three major shortcomings of research on knowledge interactions. First, the lack of established terminology hampers comparisons between the results of various research efforts. The most widely used term 'knowledge spillovers' literally implies flows from one knowledge holder/provider to several knowledge recipients, which is not always the case. Using the words *transfer* and *flow* suggests that a given piece of knowledge literally leaves the holder of the knowledge, which is not true in many cases.

Second, the vast majority of research is not focused on assessing the magnitude of real knowledge spillovers. Hence, it must be argued that only an indirect link exists between the investigated channels of knowledge interactions and real knowledge flows. Academics investigate potential channels of knowledge flows and study paper trails, not real flows (Krugman 1991) and their influence on the performance of companies and economic growth. The majority of research papers treat knowledge channels as taken for granted, as they are surely used to share knowledge. A tie or link between companies is treated as a knowledge interaction. For instance, Plum and Hassink (2013) discuss the low number of identified ties meant as knowledge flows. However, Bienkowska, Lundmark and Malmberg (2011) suggest that labour mobility does not always imply knowledge transfer.

The importance of different knowledge channels has been studied by Tödtling, Lengauer and Höglinger (2011). Their research reveals that universities and research organizations are the most important sources of knowledge for the ICT sector in the Austrian cities of Vienna and Salzburg. When it comes to the most crucial channels of knowledge interactions, R&D collaboration and informal

---

5 It may be argued that this is the employee who can value the knowledge, not a manufacturing plant.

contacts are equally important. Hence, in terms of the three channels identified by Martin and Moodysson (2011, 2013), collaboration has the largest significance.

Third, research methods and indicators need more rigorous treatment. Sotarauta, et al. (2011) correctly claim that quantitative studies do not reveal how knowledge is created or what are the main channels and geographical directions of knowledge flows. Most studies identify only the main linkages. The same approach is presented by Bienkowska, Lundmark and Malmberg (2011: 24), who argue that 'a combination of qualitative and quantitative data and analysis is particularly suitable when trying to assess, within the same space-time context, both overall patterns of labour mobility and individual motives and attitudes of central actors towards mobility'. This is why in their empirical analysis, Bienkowska, Lundmark and Malmberg (2011: 24) use two distinct sources of information: (1) statistical data at the level of individuals, (2) interviews with company representatives.

The next commonly criticized set of indicators is based on patent citations. It may be argued that patent citations include 'reviewer-added noise' in the context of citations added by a reviewer representing no new knowledge flows. Even if Jaffe, Trajtenberg and Fogarty (2000) argue that the majority of citations are the result of knowledge interactions – still not all citations reflect these flows.

Summary – Challenges for Future Research

In light of the above mentioned shortcomings and weaknesses, certain challenges for future research need to be discussed. First, the accepted terminology should be better defined and differentiated. In cases with no closed circuits of knowledge circulation, knowledge spillovers and diffusion are the best terms to use in the vast majority of cases. However, in order to avoid misunderstandings in the case of so-called closed triangles and chains of collaboration, it would be better to discuss knowledge interactions divided into direct (collaboration, mobility) and indirect types (monitoring).

Second, the extent to which each identified knowledge transmission channel reflects real knowledge flows deserves the attention of researchers. While collaboration is widely studied – even too widely, as in the case of patent citations – the role of informal relationships in knowledge interactions is often neglected. Tödtling, Lengauer and Höglinger (2011) argue that in addition to contract research and R&D collaboration, informal contacts are crucial in knowledge transfer in ICT firms. Mobility should also be more widely studied in economic geography. Finally, the role of monitoring as a third group of knowledge interactions is limited.

Third, it must be argued that quantitative research only leads to a rough approximation of knowledge interactions, so it must be enriched by survey studies. Only an integrated approach (mixed methods) allows to correctly capture knowledge interactions. This is in line with the argument of Schiller and Revilla Diez (2013: 1331), who claim that in the majority of cases 'the qualitative approach […] should be complemented with quantitative analysis' Bienkowska,

Lundmark and Malmberg (2011) correctly point out that empirical research is typically based on either statistical data or in-depth interview studies. It seems that the best solution should be to complement quantitative research (for example social network analysis) with qualitative input.

References

Abel, J.R. and Deitz, R. 2012. Do colleges and universities increase their region's human capital? *Journal of Economic Geography*, 12, 667–91.

Agostino, L.M.D., Laursen, K. and Santangelo, G.D. 2013. The impact of R&D offshoring on the home knowledge production of OECD investing regions. *Journal of Economic Geography*, 13, 145–75.

Agrawal, A., Cockburn, I. and McHale, J. 2006. Gone but not forgotten: knowledge flows, labor mobility, and enduring social relationships. *Journal of Economic Geography*, 6, 571–91.

Alazzawi, S., 2011. Multinational Corporations and Knowledge Flows: Evidence from Patent Citations. *Economic Development and Cultural Change*, 59(3), 649–80.

Alkay, E. and Hewings, G.J.D. 2012. The determinants of agglomeration metropolitan area. *Annals of Regional Science*, 48, 225–45.

Almeida, P. and Kogut, B. 1999. Localization of knowledge and the mobility of engineers in regional networks. *Management Science*, 45, 905–18.

Asheim, B.T., Smith, H.L. and Oughton, C. 2011. Regional Innovation Systems: Theory, Empirics and Policy. *Regional Studies*, 45(7), 875–91.

Basile, R., Capello, R. And Caragliu, A. 2012. Technological interdependence and regional growth in Europe: Proximity and synergy in knowledge spillovers. *Papers in Regional Science*, 91(4), 697–722.

Bathelt, H. and Schuldt, N. 2008. Between Luminaires and Meat Grinders: International Trade Fairs as Temporary Clusters. *Regional Studies*, 42(6), 853–68.

Bienkowska, D., Lundmark, M. and Malmberg, A. 2011. Brain circulation and flexible adjustment: Labour mobility as a cluster advantage. *Geografiska Annaler Series B: Human Geography*, 93(1), 21–39.

Boschma, R. and Frenken, K. 2011. The emerging empirics of evolutionary economic geography. *Journal of Economic Geography*, 11(1), 295–307.

Brown, W.M. and Rigby, D.L. 2010. *Marshallian localization economies: where do they come from and to whom do they flow?* Paper presented at the DIME Workshop 'Industrial Dynamics and Economic Geography' (Utrecht, The Netherlands).

Buenstorf, G. and Geissler, M. 2010. The origins of entrants and the geography of the German laser industry. *Papers in Regional Science*, 90(2), 251–71.

Burris, V. 2005. Interlocking Directorates and Political Cohesion among Corporate Elites. *American Journal of Sociology*, 111(1), 249–83.

Campbell, C., Nair, Y., Maimane, S., Sibiya, Z. and Gibbs, A. 2011. 'Dissemination as intervention': Building local HIV Competence through the report back of research findings to a South African rural community. *Antipode*, 44(3), 702–24.

Croce, G. and Ghignoni, E. 2013. Employer-provided training and knowledge spillovers: evidence from Italian local labour markets. *Regional Studies*, 46(3), 339–53.

Desrochers, P. and Leppala, S. 2011. Opening up the 'Jacobs Spillovers' black box: local diversity, creativity and the processes underlying new combinations. *Journal of Economic Geography*, 11, 843–63.

Eriksson, R.H. 2011. Localized spillovers and knowledge flows: how does proximity influence the performance of plants? *Economic Geography*, 87(2), 127–52.

Expósito-Langa, M., Molina-Morales, F.X. and Capó-Vicedo, J. 2011. New product development and absorptive capacity in industrial districts: a multidimensional approach. *Regional Studies*, 45(3), 319–31.

Feldmann, M.P. 2003. Location and innovation, in *The Oxford Handbook of Economic Geography*, edited by G.L. Clark, M.P. Feldman and M. Gertler. New York: Oxford University Press.

Fischer, M.M. 2011. A spatial Mankiw-Romer-Weil model: theory and evidence. *Annals of Regional Science*, 47, 419–36.

Fritsch, M. and Slavtchev, V. 2011. Determinants of the efficiency of regional innovation systems. *Regional Studies*, 45(7), 905–18.

Gallie, E.-P. 2009. Is geographical proximity necessary for knowledge spillovers within a cooperative technological network? The case of the French biotechnology. *Regional Studies*, 43(1), 33–42.

Grabher, G. 1994. The Disembedded regional economy: the transformation of east german industrial complexes into western enclaves, in *Globalization, Institutions, and Regional Development in Europe*, edited by A. Amin and N. Thrift. Oxford: Oxford University Press, 177–95.

Hardy, J. 1998. Cathedrals in the desert? Transnationals, corporate strategy and locality in Wrocław. *Regional Studies*, 32(7), 639–52.

Heebels, B. 2013. Place-making in publishing: Dutch trade book publishers and the meaning of place in establishing reputation and trust. *Geoforum*, 45, 248–58.

Hennemann, S. 2011, The role of transnational corporation in the Chinese science and technology. *Erdkunde*, 65(1), 71–83.

Huber, F. 2012. Do clusters really matter for innovation practices in Information Technology? Questioning the significance of technological knowledge spillovers. *Journal of Economic Geography*, 12(1), 107–26.

Huggins, R. and Kitagawa, F. 2013. Regional policy and university knowledge transfer: perspectives from devolved regions in the UK. *Regional Studies*, 46(6), 817–32.

Jaffe, A.B, Trajtenberg, M. and Fogarty, M.S. 2000. Knowledge spillovers and patent citations: evidence from a survey of inventors. *American Economic Review*, 90(2), 215–18.

Jaffe, A.B. and Henderson, R. 1993. Geographical localization of knowledge spillovers as evidenced by patent citations. *Quarterly Journal of Economics*, 108, 577–98.

Jordaan, J.A. and Rodriguez-Oreggia, E. 2012. Regional growth in Mexico under trade liberalisation : how important are agglomeration and FDI ? *Annals of Regional Science*, 48, 179–202.

Kemeny, T. 2011. Are international technology gaps growing or shrinking in the age of globalization? *Journal of Economic Geography*, 11(1), 1–35.

Krugman, P. 1991, *Geography and Trade*. Cambridge, MA: MIT Press.

Lin, G.C.S., Wang C.C., Zhou Y., Sun Y. and Wei, Y.D. 2011. Placing technological innovation in globalising China: production linkage, knowledge exchange and innovative performance of the ICT industry in a developing economy. *Urban Studies*, 48(14), 2999–3018.

Martin, R. and Moodysson, J. 2011. Innovation in symbolic industries: the geography and organization of knowledge sourcing. *European Planning Studies*, 19(7), 1183–203.

Martin, R. and Moodysson, J. 2013. Comparing knowledge bases: on the geography and organization of knowledge sourcing in the regional innovation system of Scania, Sweden. *European Urban and Regional Studies*, 20(2), 170–87.

Micek G., Działek J. and Górecki J. 2011. The Discourse and Realities of Offshore Business Services to Kraków. *European Planning Studies*, 9, 1651–68.

Mitton, C., Adair, C., McKenzie, E., Patten, S. and Perry, B. 2007. Knowledge transfer and exchange: review and synthesis of the literature. *The Milbank Quarterly*, 85(4), 729–68.

Morris, J. 1992. Flexible internationalisation in the electronics industry: implications for regional economies. *Environment and Planning C*, 10, 407–21.

Mulrennan, M.E., Mark R. and Scott, C.H. 2012. Revamping community-based conservation through participatory research. *The Canadian Geographer*, 56(2), 243–59.

Ó Huallachain B. and Lee, D.-S. 2011. Technological specialization and variety in urban invention. *Regional Studies*, 45(1), 67–88.

Plum, O. and Hassink, R. 2013. European urban and regional studies firms. *European Urban and Regional Studies*, 20(2), 206–26.

Potter, A. and Watts, H.D. 2011. Evolutionary agglomeration theory: increasing returns, diminishing returns, and the industry life cycle. *Journal of Economic Geography*, 11(2), 417–55.

Raymond, C.M. and Robinson, G.M. 2013. Factors affecting rural landholders' adaptation to climate change: Insights from formal institutions and communities of practice. *Global Environmental Change*, 23(1), 103–14.

Rychen, F. and Zimmermann, J-B. 2008. Clusters in the global knowledge-based economy: knowledge gatekeepers and temporary proximity. *Regional Studies*, 42(6), 767–76.

Scherngell, T. and Barber, M.J. 2011. Distinct spatial characteristics of industrial and public research collaborations: evidence from the fifth EU Framework Programme. *Annals of Regional Science*, 46(2), 247–66.

Schiller, D. and Revilla Diez, J. 2013. The impact of academic mobility on the creation of localized intangible assets. *Regional Studies*, 46(10), 1319–32.

Schuldt, N. and Bathelt, H. 2011. International trade fairs and global buzz. Part II: Practices of global buzz. *European Planning Studies*, 19(1), 1–22.

Sonn, J.W. and Lee, D. 2012. Revisiting the branch plant syndrome: review of literature on foreign direct investment and regional development in Western advanced economies. *International Journal of Urban Sciences*, 16(3), 243–59.

Sotarauta, M., Ramstedt-Sen, T., Seppänen, S.K. and Kosonen, K.-J. 2011. Local or digital buzz, global or national pipelines: patterns of knowledge sourcing in intelligent machinery and digital content services in Finland. *European Planning Studies*, 19(7), 1305–30.

Tabata, M. 2013. The absorption of Japanese engineers into Taiwan' s TFT-LCD industry : globalization and transnational talent diffusion. *Asian Survey*, 52(3), 571–94.

Tödtling, F., Lengauer, L. and Höglinger, C. 2011. Knowledge sourcing and innovation in 'Thick' and 'Thin' regional innovation systems – comparing ICT firms in two Austrian regions. *European Planning Studies*, 19(7), 1245–76.

Tödtling, F., Skokan, K., Höglinger, Ch., Rumpel, P. and Grillitsch, M. 2013. Innovation and knowledge sourcing of modern sectors in old industrial regions: comparing software firms in Moravia-Silesia and Upper Austria. *European Urban and Regional Studies*, 20(2), 188–205.

Usai, S. 2011. The geography of inventive activity in OECD regions. *Regional Studies*, 45(6), 711–31.

Wang, C.C. and Lin, G.C.S. 2013. Emerging geography of technological innovation in China's ICT industry: region, inter-firm linkages and innovative performance in a transitional economy. *Asia Pacific Viewpoint*, 54(1), 33–48.

Weiss, K., Haman, M., Kinney, M. and Marsh, H., 2012. Knowledge exchange and policy influence in a marine resource governance network. *Global Environmental Change*, 22(1), 178–88.

Wenting, R., Atzema, O. and Frenken, K. 2011. Urban amenities and agglomeration economies? The locational behaviour and economic success of Dutch fashion design entrepreneurs. *Urban Studies*, 48(7), 1333–52.

PART II
Large Firms and FDI: Global–Local Dichotomy

Chapter 4
Geographical Embeddedness of Large Corporations in Turkey

Aksel Ersoy

A fundamental issue around large corporations is to understand how they contribute to the formation of national and global economies. Starting from Perroux (1955), who viewed growth within the network of industries, the network linkages amongst enterprises have been considered both at the regional and national level through input–output relations, and at the international level through cross-border investment by transnational corporations (TNCs) (Myrdal 1957, Hirschman 1958, Perroux, 1973). However, these ongoing interactions started to constrain institutions as they became no longer 'independent' (Granovetter 1985: 482). Granovetter (1985) explained the concept of embeddedness in terms of the role of social relations in economic transactions and this has given rise to a powerful model of local economic growth that draws on a range of complementary literatures on new industrial spaces, learning regions, innovative milieu and regional innovation systems, clusters, and the creative class (for example Storper 1997, Braczyk et al. 1998, Porter 1998, Florida 2002, MacKinnon, Cumbers and Chapman 2002, Cumbers, MacKinnon and Chapman 2003). The aim of this chapter is to study the geographical agglomeration of the top 500 large corporations in Turkey. It argues that local embeddedness of those corporations leads to an institutional thickness that is thought to be one crucial success factor for regions in a globalizing economy.

After the emergence of 'embeddedness' in economic behaviour, the relationship between economic activity and spaces have been widely employed by economic geographers. Referring back to the work of Polanyi (1944) and Granovetter (1973, 1985), geographers have used the concept from a distinct 'spatial' point of view to explain the economic success of regions built around industrial clusters. While some studies place embeddedness central in theories of regional economic development (Dicken and Malmberg 2001, Cooke 2002), others explain global–local dichotomy (Amin and Thrift 1994, Conti 1997, Martin and Sunley 2001). They highlight the importance of local TNC embeddedness as they generate spillover effects for local development. This has created an interest in examining the geographical dimensions of corporations. Studies analyse the impacts of corporations' engagement with regions and localities in processes of economic development (Storper 1997, Oinas 1997, Yeung, Poon and Perry 2001). They argue that local economies are embedded in large corporations as they are the key points for market exchange.

However, this situation is different for those externally owned companies. A considerable body of literature has shown that externally owned corporations fail to generate both local and regional success (Phelps et al. 2003, Pavlínek 2004, van Egaraat and Jacobson 2005). This failure not only affects the performance of business services but also influence the geographical distribution of high skilled and knowledge workers (Guzik and Micek 2008). What is more is that these studies have focussed on economically advanced and technologically innovative developed countries. Even though recent studies explain the increasing embeddedness of TNCs (Domański 2004, Domański and Gwosdz 2009), more research needs to be undertaken. Many factors such as quality and quantity of local suppliers, the role of export platforms, and lack of intellectual property rights (IPRs) affect the relationship between TNCs and domestic enterprises in developing countries (Yeung and Li 2000).

As a developing country, Turkey is an interesting case study to observe the relationship between TNCs and local enterprises. First of all, social networks and personal relations matter in the Turkish culture. A close relationship between customers and firms accelerate the creation of network linkages amongst enterprises. Bayirbag (2010) explains the success of Gaziantep, in terms of the number of business associations and their engagement in the locality which have been created on social relations. This means formal and informal linkages between companies and stakeholders have the potential to enhance local creativity which may have strong influence on attracting TNCs. The second issue is about the structure of the enterprises in Turkey. Like any other developing country, state owned enterprises (SOEs) have dominated the Turkish economy for many decades. After Turkey's encounter with the neoliberal reforms in the 1980s, privatization has become a major issue in Turkey's economy. This has attracted various TNCs as they have become partners with SOEs. Therefore, TNCs have been gradually embedded with local partners emerging new economic spaces of innovation and creativity in Turkey. Finally, located at the intersection of Europe, the Middle East, the Balkans and the former Soviet Union, the Turkish economy has registered itself on the world economic scale as an exceptional emerging economy, the 16th largest economy in the world and the 6th largest economy when compared with the EU countries, according to GDP figures (at PPP) in 2010 (*Invest in Turkey* 2011). Turkey has a rapidly growing economy built especially on manufacturing. Productivity is increasing within the national economy and the style of Turkey's economic growth has been likened to that of China and India (Rodrik 2010). At the same time, Turkey is positioning itself and preparing for entry into the European Union. Therefore, Turkey's geopolitical position and economic promises provide a potential market for many TNCs which are linked to geographical, institutional and social characteristics and changes in Turkey.

However, the global economic environment with which Turkey's regional economic policy must cope is also changing rapidly. Global interdependence is increasing on a massive scale orchestrated through TNC-driven value chains, outsourcing, and evolving styles of value chain management. The 'financialization' of the corporations driving these chains has seen their focus shifted towards the creation shareholder value, while externalizing the risks of production. Increasingly, they are owners of assets and orchestrators of IPRs, technology and financial flows from which they can strip profits while externalizing and outsourcing production to contract manufacturers. And they are aided in their operations by financial institutions, especially the so-called 'clever' and formalized speculation of hedge funds and financial intermediaries (Taylor and Ersoy 2011). Turkey's manufacturers, unfortunately, tend to operate in the lower ranges of these corporate controlled value chains which take part in the consumer debt of the triad economies of North America, Europe and Japan at the international level. Large investments in Turkey's less-developed urban areas could lead to the substantial restructuring of the hierarchy of the Turkish provinces which would have an effect on the reduction of the regional disparities.

This chapter examines the geography of the *Top 500* large corporations in the Turkish urban system, giving consideration to their locations, performance and concentrations. It argues that not only the spatial organization of corporations remains important in urban systems but also they are linked to the dynamics of local and regional economies within developing countries. The chapter begins by considering the literature on geographical embeddedness of corporations. It then considers the structure of Turkey's largest 500 listed firms, focusing on the sectoral distribution of companies and profits at both sectoral level and provincial level. The following section deals with the geography of large corporations and the degree to which agglomeration is evident in the Turkish provinces.

Geographical Embeddedness of Corporations

Built on Granovetter's (1985) concept of embeddedness, a range of theories of economic growth have developed that move beyond the anonymity of pure market mechanisms to emphasize economic life and commercial transactions built on social interconnections. The term 'embeddedness' here refers to the fact that economic development is affected by the network of social relations (Grabher 1993: 4). Hess (2004) distils three major dimensions of what compromises embeddedness and who is embedded in what. They are societal embeddedness, network embeddedness and territorial embeddedness (Hess 2004: 176–7). Societal embeddedness signifies the importance of a cultural and political background influencing and shaping the action of actors within society (Gertler 2001, Harzing and Sorge 2003). The main argument behind this type of embeddedness is to emphasize the history of social networks and the cultural imprint or heritage of actors that influence economic behaviour both at the national and international level. Network embeddedness describes the network of actors and the structure of relationships among a set of individuals and organizations regardless of their country of origin or local anchoring in particular places (Hess 2004: 177). This type of embeddedness is regarded as the product of a process of trust building between network agents which is essential for a sustainable relationship. Territorial embeddedness considers the extent to which an actor is spatially placed in particular territories. According to Hess (2004: 177), economic actors become embedded, or in some cases become constrained by, the economic activities and social dynamics that already exist in space. For example, the location of TNCs might generate a new local or regional network of economic and social relations. Embeddedness, then, may become a key element in regional economic growth and in capturing global opportunities (Harrison 1992, Amin and Thrift 1994). Because this chapter focuses on the distribution of the top 500 large corporations in Turkey, the nature of the embeddedness concept is more concerned with territorial embeddedness.

In the literature, the crucial part of the geographical embeddedness of corporations debate has explored how the spatial concentrations of corporations contribute to the formation of national or international management centres and the role that these play restructuring national and global economies (Goodwin 1965, Taylor and Thrift 1982, Tonts and Taylor 2010). While the work of Taylor (2004, 2005, 2007) on world cities has shown how corporate activities are embedded within the global command centres of national economies, other studies have explained the importance of the spatial organization of corporate headquarters in shaping local and regional development (Holloway and Wheeler 1991, Chan, Gau and Wang 1995, Carroll 2001, Klier 2006, Quark 2007). Tonts and Taylor (2010) have recently argued seven underlying advantages that shape the distribution of corporations within national economies. They are the role of urban management centres as sources of strategic and specialist information; the importance of large metropolitan areas as transport and communication hubs; the co-location of proximity of other enterprisers; the concentration of skilled and professional people; the concentration of business service firms offering a wide range of specialist activities and solutions; the foci of social and class-based support structures that contain privileged elites for corporate board members; and the outcome of processes of path-dependent local and regional development (Tonts and Taylor 2010). Even though these advantages exist within national economies, the social, political, and economic circumstances in which corporations are located change very rapidly and hence advantages that were once in place would be displaced in time. In some cases, however, the changes would contribute to the reproduction of place through the presence of a specialist labour market, government, research institutions access to natural resources, or particular policy frameworks (Tonts and Taylor 2010).

However, some authors remain sceptical about those externally owned companies. Pavlínek (2004) argues that the effects of foreign direct investments in Central Europe have been uneven both sectorally and geographically. He identifies negative effects of FDI on local and regional economic development and argues that the CE economies, dependent on FDI, and are becoming more vulnerable to plant closures if local currencies appreciate rapidly or local wages increase substantially. Van Egaraat and Jacobson (2005) show that multinational plants of the software industry are weakly embedded in the regional economies of

Ireland and Scotland. They argue that a strategy for the microcomputer industry is becoming increasingly unstable in Ireland and Scotland. Ersoy (2012) also shows how foreign owned businesses operate in Turkey and how people achieve local growth and local knowledge transfer out of the branch plant economy. He argues that in Turkey, the branch plant economy is either externalized through sub-constructers working for overseas firms, or it is operated through the main headquarters which are located in big metropolitan cities. Although knowledge creation is an essential element of economic growth, the empirical analysis has shown that the knowledge creation variable is negatively related with economic growth in Turkey (Ersoy and Taylor 2012).

These studies have shown that large corporations have become a major part of our society. Tonts and Taylor (2010) explain this reality at two different levels. The first is global-level processes that focus on the level of economic integration of places. The topics cover the emergence of global centres (Taylor, Ni and Derudder 2009, Sassen 2006); the division of production and control locations (Phelps 1993, Bartlett and Hedlund 1996); and the multi-national corporations (Carroll 2007). The second focuses on the sub-national level by understanding processes of economic development in which corporations engage with regions and localities (Storper 1997, Oinas 1997, Yeung et al. 2001). Similarly, Scott (1997: 26) explains three kinds of economic power that the corporate form of economic organization entails in society: *strategic, operational, and allocative.* Strategic power occurs at the level of structural decision making and concerns the determination of basic long-term goals and the adoption of initiatives to realize those goals. Operational power involves the actual implementation of corporate strategy within the head office and in sub-ordinate offices, subsidiaries, and plants. Finally, there is the allocative power wielded by financial institutions, whose collective control over the availability of capital "gives them the power to determine the broad conditions under which other enterprises must decide their corporate strategies" (Scott, 1997: 139).

However, although large corporations are an essential element of economic growth, the empirical analysis has shown that the power of large corporations is not significant with economic growth in Turkey (Ersoy and Taylor 2012). They suggest a very distinctive form of production-based regional economic growth across Turkey's provincial economies built on: linked small firm manufacturing, local human capital, local institutional support, and access to intermediate goods markets. It is an interpretation that is consistent with Turkey's manufacturing and production firms being subordinated within value chains dominated by large, and in many cases foreign-owned, corporations. Ersoy (2012) also investigates the power of large corporations to enhance or retard local growth in the Turkish context. He develops the surrogate from the 2008 'Top 500 industrial enterprises' list of ISO (2009) by assigning each business to the head office province (Figure 4.1) and the net production profit for each corporation has been added up to find the total net profit of production for a particular province (Figure 4.2). It is important to acknowledge that the analysis does not make a distinction between local and foreign owned corporations due to the limitation of the data, even though this difference might have different implications for geographical embeddedness. To estimate the impact of corporations in that province, the net productive profit of each corporation has been assigned to the province within which it is headquartered and standardised across provinces as 'production profit per resident person in a particular province'.

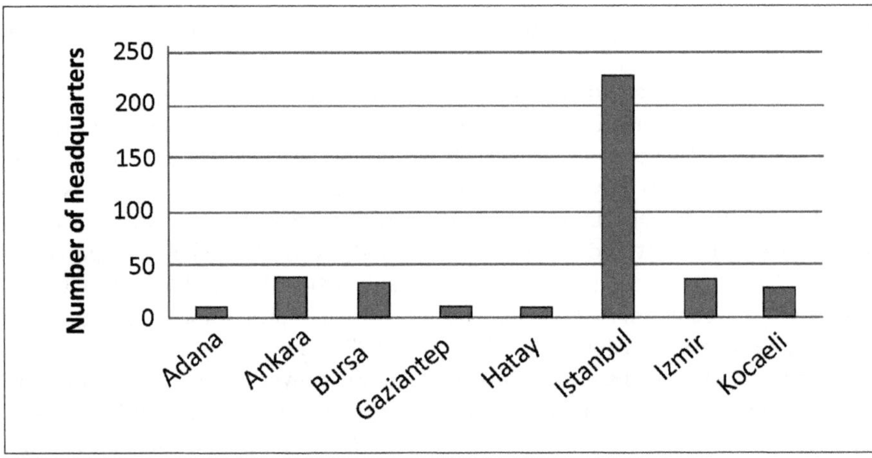

Figure 4.1 Location of company headquarters in Turkey
Source: own elaboration based on ISO (2009).

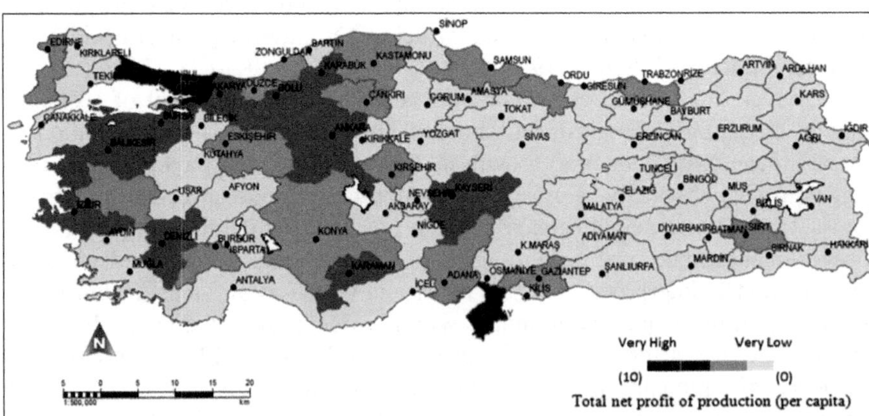

Figure 4.2 Total net profit of production (per capita) of Top 500 industrial enterprises in Turkey
Source: own elaboration based on ISO (2009).

According to the result of his analysis (Figure 4.2), some big provinces such as Istanbul, Izmir and Ankara still remain significant in the sense that the total net production profit of corporations in that province is above average (Ersoy 2012). However, some provinces such as Balikesir, Denizli, Hatay and Karaman appear significant as well. These are important findings for the localities that have not been significant or explored in detail. As Guzik and Micek (2008) suggest, the geographical distribution of large corporations not only affects the performance of business services but also influence the geographical distribution of high skilled and knowledge workers. In fact, analysis shows that there is a significant difference in the intensity of knowledge workers in metropolitan areas, mainly in Istanbul, Ankara and Izmir as these cities have more connections with the global economy and hence they are more closely associated with economies of scale and agglomeration economies (Ersoy 2012). These cities have become the attraction areas for high skilled and well paid knowledge workers. The rest of the concentrations can be interpreted as intermediates and bridges in the knowledge economy. Not only do they play significant local and regional roles in Turkey, they also link Turkey and its regional economies into the global economy. Denizli and Kayseri are very good examples for this global connection. This can be explained by the fact that endogenous potentials of these provinces are much more significant than other provinces (Eraydin 2002). Therefore, these provinces might constantly reinvent themselves, moving from one field of specialization to another. Skills may well be a crucial part of this reinvention as skilled workers react more speedily to painful economic shocks (Welch 1970).

The Changing Economic Landscape in Turkey

Understanding the geographical embeddedness of corporations in Turkey is much more complicated. Heper (1991) argues that the economic policy process in Turkey has played an important role in shaping the business outlook of the country. Even though the outcome of path dependency is still significant in the successful Turkish cities (Eraydin 2002), the Turkish state intervention has resulted in ambiguities concerning the legitimacy of business activities. The obvious example for this is the economic performances of the Turkish provinces. The provinces located in the eastern parts of Turkey are known to be lagging behind in economic and social terms. Studies point out that the geographical location of provinces influences the level of income and education in the favour of the western regions (Celebioglu and Dall'erba 2009, Gezici and Hewings 2004). Similarly they show that inequalities in salaries and migration from east to west increase the rate of urbanization (Elveren and Galbraith 2008, Kirdar and Saracoglu 2007) which is one of the main problems of Turkey's seven regions.

Throughout the years, the policies and practices of Turkish local and regional economic development have changed to balance the divergence between the western and eastern regions of Turkey. However, Turkish regional imbalances have been the subject of Turkish reality for many years. For instance, Ertugal (2005) explained that redistribution policies did not necessarily achieve the aim of increasing cohesion across Turkish regions. Gezici and Hewings (2004) point out the importance of functional regions of Turkey and conclude that the regions that receive most of the funds did not grow faster than the more developed regions of the West. To explore the dynamics and structures of Turkish regions, it is crucial to understand the changes and characteristics of the policies and practices of Turkish local and regional economic development.

After the Second World War, infrastructure provision was the main concern addressed by Turkish regional policy. Agricultural production started to lose its importance and problems in the traditional heavy industries led to increasing unemployment rates in the regions. This was followed by massive immigration flows to city centres. Increasing density in the centres was becoming a major problem and attempts at solving this problem involved providing infrastructure which was mainly provided by the state. Since the nation state was the main actor in the development process, it was expected that it should manage the economy, be responsible for the inequalities and be in charge of the allocation of resources (Tekeli 2004). Therefore, the policies between 1945 and 1970 were mainly structured according to state interventions.

The economic crisis that was related to the two oil crises of the 1970s raised questions regarding the impact of state interventions. Centralist regional policies started to lose their importance and the nation state could no longer maintain their regional policies due to a lack of resources (Eraydin 2002). Therefore, the policy focus shifted towards privatization, deregulation and the liberalization of markets. Regional policy moved progressively down the policy agenda, downgrading the policy goals of reducing disparities and promoting regional convergence (Bachtler, Wishlade and Yuill 2001). Nevertheless, the period between 1970 and 1990 also encouraged regions that were competitive and had learning capacity. In this respect, a new regionalism based on endogenous development was encouraged during this period. So there was a move from the region to the local which had its own characteristics, accumulation and endogenous potential and this was implemented as a new policy (Eraydin 2002).

In the 1980s, this transformation became more significant in the sense that the region was identified as a combination of local potential, networks and institutions. With this new transformation, regions started to be defined in accordance with various networks in the world (Eraydin 2004). Since central government started to lose its power on regional and industrial policy, regional and local aspects became crucial in regional development. Therefore, economic development policies started to be considered at the local level by empowering the localities. This encouraged the participation of local governments in the decision making process and to develop their own solutions when they faced problems. That kind of orientation highlighted the importance of endogenous development, entrepreneurship and the SMEs.

After the 1980s, globalization weakened the border concept of nations and regional competitiveness became crucial in economic development. However, the globalization process did not always bring success to regions. Some regions could not adapt to changing economic conditions and this resulted in huge regional inequalities. While some regions enjoyed the competitiveness of their region thanks to their regional potential, some regions started to lag behind the national economy. To be able to cope with the difficulties of regional competitiveness, new strategies and precautions were defined by the nations, the OECD and the EU. In terms of regional growth, endogenous development replaced external demand and it became part of regional growth after the 1970s. There was a shift in dynamics of regional growth from capital accumulation to human capital between the 1950s and the 1990s which was then dominated by technology and innovation. Path dependency and networks became important elements of Turkey's comparative advantage. The centralist nation State was no longer powerful in regional policy. Decentralization of power became crucial. In terms of the policy tools that have been utilized in Turkey's regional policy, it was realized that direct investment was no longer valid for regional development. To increase the comparative advantage of the nation state, clusters and innovative regions emerged. To be able to access the wider parts of the global production chain, networks became crucial and they have been located at the heart of policy tools. While the global economic crisis that happened during the last decade had a clear effect on the Turkish economy, it is too early to determine if it is a major shift in the nation's changing corporate landscape. The notion of path dependency is crucial to the concept of embeddedness because large corporations are able to generate structural changes and modify the behaviour of the system. According to Sabel and Zeitlin (1997), path dependency is a result of both self-reinforcing mechanisms and the reflexivity of actors. Actors can alter the direction of the paths they follow and reshape new paths.

For example, in the studies of corporate control in Turkey, Aytac and Sak (2000) point to the role of the dominant shareholders, for instance families, as they controlled more than 50 per cent of the voting rights. Similarly, Demirag and Serter (2003) examine the ownership structure of Turkish companies and they find out that a majority of them are owned or controlled by families. They show that ownership amongst Turkish companies is highly concentrated, with 38 per cent of the largest 100 traded companies owned by a single large shareholder (Demirag and Serter 2003). This picture is also very similar within the distribution of the top 500 enterprises. The Istanbul Chamber of Industry[1] (ISO) produces a list of the 'Top 500 industrial enterprises' every year. Although the top 10 companies represent almost 28 per cent of total production or sales in 2008, the share of five Koc[2] companies among the top 10 is around 45 per cent (Table 4.1). What is more

1 The data can be accessed through the website www.iso.org.tr.

2 Koc Holding is Turkey's dominant industrial conglomerate. Koc Holding has five companies in the Top 10 Companies list: Tupras–Turkiye Petrol Refinerileri, Ford Otomotiv Sanayi AS, Tofas Turk Otomobil Fabrikasi, Arcelik and Aygaz.

striking is that Tupras (Turkey's largest industrial enterprise based on crude oil processing) alone represents almost per cent 40 per cent of the top 10 companies' sales. Tupras is four-and-a-half times larger than the second biggest firm. Another observation is that Koc companies compete with each other at the top of the industrial rankings. The figures are some serious signs of a rising concentration, or oligopoly structure, in Turkish industry.

Table 4.1 Top 10 Companies in the 'Top 500 industrial enterprises' list

Rank	Companies	Profit (TRY)
1	Tupras-Turkiye Petrol Rafinerileri A.S.	27,732,867.295
2	EUAS Elektrik Uretim A.S. Genel Mudurlugu	6,249,112.724
3	Ford Otomotiv Sanayi A.S.	6,006,491.811
4	Eregli Demir ve Celik Fabrikalari T.A.S.	5,014,572.054
5	Oyak-Renault Otomobil Fabrikalari A.S.	4,710,974.763
6	Tofas Turk Otomobil Fabrikasi A.S.	4,184,361.976
7	Arcelik A.S.	4,068,892.569
8	Icdas Celik Enerji Tersane ve Ulasim San. A.S.	3,828,300.738
9	Habas Sinai ve Tibbi Gazlar Istihsal Endustrisi A.S.	3,476,676.147

Source: ISO (2009).

The overall structure in corporate concentration can be conceptualized as the outcome of broader macroeconomic and political circumstances such as the emergence of a new international division of labour, the growth of the service economy and the increasing fiscal nature of companies (Taylor and Thrift 1980). In terms of macroeconomic circumstances, the most significant was the onset of the years between 1999 and 2001, which contributed to low levels of economic growth, high unemployment and inflation. Fortunately, a successful IMF-sanctioned economic restructuring brought foreign investment to the country. The banking sector has benefitted from the recent developments because larger strategic investments have made large investments in Turkey (Kaymak and Bektas 2008). This stimulated the notion of the segmented economy that seek to describe a shift in the nature of market economies from one in which small firms and individual producers were dominant to an economy in which large firms play a significant role. Therefore, the economies of scale that accrue to the large firm enable a division of labour within the large firms. This process can be considered as the onset of deindustrialization in Turkey and the gradual shift of some manufacturing and production based industries offshore, as companies sought relief from relatively high taxes and high labour costs.

The response of Turkish governments to the economic problems was based on widespread liberalization of the economy such as the liberalization of the banking and finance sector, a relaxation of foreign direct investments and privatization (see Sahin 2010, Onis 2011). Turkey has generated a sum of $41.98 billion in privatization revenues between 1985 and 2010 in which a major part of it has been accrued after 2001. The government declared its commitment for the privatization of the state owned enterprises (SOEs) in the Emergency Action Plan in 2003. The government adopted the Turkish Privatization Strategy Plan on 18 May 2003 and emphasized the privatization generated economic growth in the future (see Privatization Strategy Plan 2003). The formal negotiations with the European Union accelerated privatization revenues in the post-2004 period (Figure 4.3).

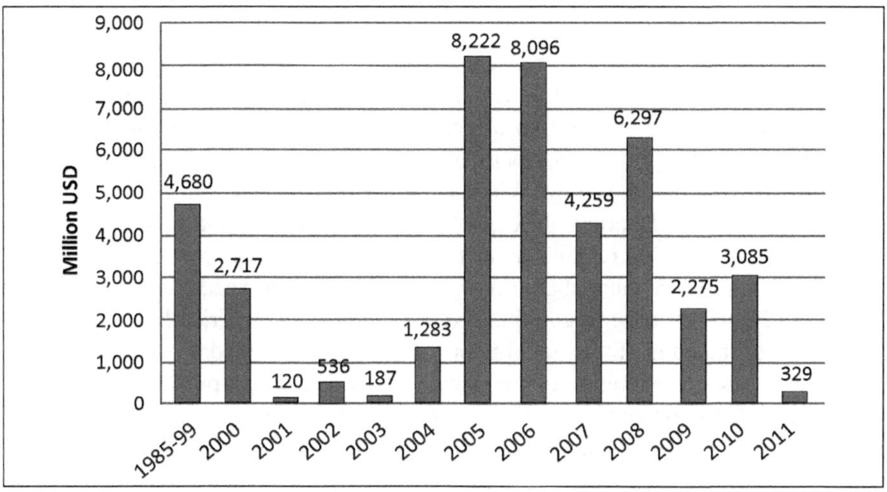

Figure 4.3 Total privatization values in Turkey (1985 onwards)
Source: own elaboration based on ISO (2009).

During the privatization period in Turkey, block sales have been the main form of transfer with 48 per cent. This is followed by public offering which is composed of sale to employees, through ordinary sale or private placement in stock exchanges and sale to investment funds; and asset sale which is the transfer of production and service units and other assets of enterprises in return for a price. The privatization strategy is adopted by State governments in a popular way; indeed, what are now some of Turkey's largest companies were formerly government-owned enterprises. Alongside this, in the last decade, the relaxation of financial and banking regulations have experienced a number of increasing mutual funds converted to listed companies. The privatization and liberalization of formerly government controlled activities led to a continuing globalization of activities and service provision. Privatization and decentralization not only allowed local private capital into large corporations, but also they generated a wave of foreign investment. This has reshaped the power relations within different sectors and changed the nature of value-added networks.

Distribution of Corporations in Turkey

The geography of large corporations in Turkey is highly concentrated around big metropolitan cities. Amongst these cities, Istanbul is pre-eminent, accounting for 227 corporations of the top 500. Ankara (41 companies), Izmir (36 companies), Bursa (32 companies) and Kocaeli (28 companies) play important second-tier roles, while Kayseri (13 companies), Adana (11 companies) and Gaziantep (10 companies) are relatively peripheralized from the Turkish corporate landscape. Nevertheless, there is a complex relationship between population concentrations and the location of these corporations. Figure 4.4 provides an indicator for this relationship. In the case of Istanbul, the concentration of company headquarters relative to population is particularly striking, with 45.4 per cent of headquarters and 23.4 per cent of the aggregate population of Turkey. While Izmir, Bursa and Kocaeli still have higher concentrations of headquarters, the capital city Ankara's concentration of headquarters relative to its population remains almost the same with 8.2 per cent. These findings also match with other international research undertaken on headquarters. Klier and Testa (2002), for instance, argue that the largest cities in America have the largest company headquarters. Tonts and Taylor (2010) show that Sydney and Melbourne have the highest concentration of company headquarters and population in Australia. Studies that focus on manufacturing clusters in Turkey also reveal that firms are localized in four metropolitan areas (Istanbul, Izmir, Ankara and Adana) and that these four regions make up nearly 73 per cent of the total manufacturing labour force (Eraydin 2002). The relationship between the concentration of company headquarters and population in Adana and Gaziantep, on the other hand, is the opposite. Hatay's situation, however, is strikingly different with 1.8 per cent of headquarters and 0.1 per cent of the aggregate population of Turkey.

A more complex relationship is evident between the geographical location of headquarters and the spatial concentration of the production profit, net value added and export figures of top 500 companies (Figure 4.5). Istanbul clearly dominates in Turkey, accounting for more than 45 per cent of headquarters, production profit, net value added and half of all exports. Kocaeli, even though the distribution of headquarters in percentages is less than Ankara (4.8 per cent), Bursa (3.6 per cent) and Izmir (5.4 per cent), outperforms all the other cities in terms of production profit. Moreover, in terms of net value added, Ankara and Kocaeli remain the only cities in which the net output has the highest value amongst other indicators. This is an important fact for companies in Ankara and Kocaeli. Added value allows those companies to market their products more successfully, emphasizing inimitability and strength of brand as opposed to a commodity. Kocaeli has also

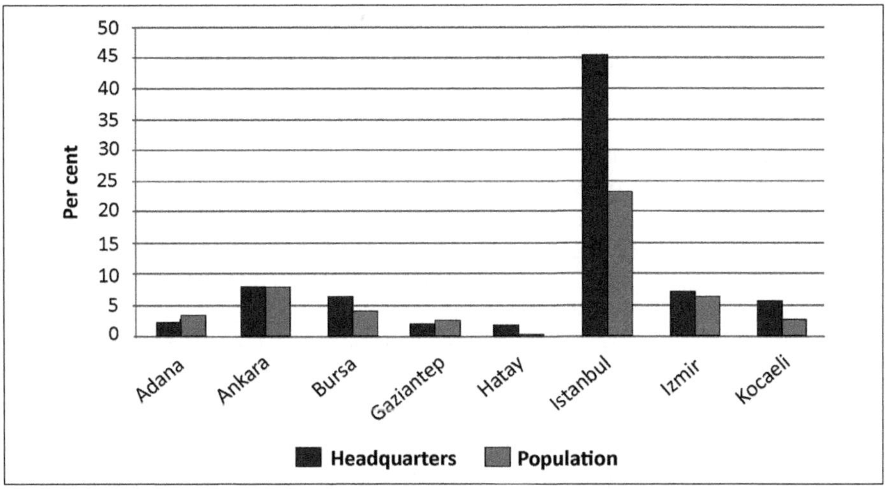

Figure 4.4 Concentrations of headquarters and population in Turkey
Source: own elaboration based on ISO (2009).

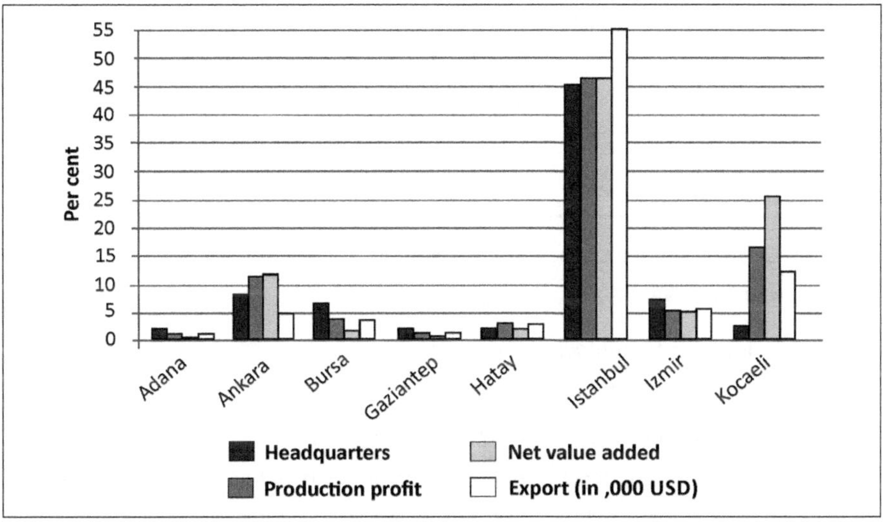

Figure 4.5 Distribution of headquarters, production profit, net value added and export in percentages
Source: own elaboration based on ISO (2009).

staggering export values with 11.8 per cent, followed by Izmir (5.4 per cent) and Ankara (4.8 per cent). The distribution of these values is in part explained by the sectoral mix and degree of specialization within Turkish cities. To explore this, the nature of concentration is investigated for each sector in Adana, Ankara, Bursa, Gaziantep, Hatay, Istanbul, Izmir and Kocaeli. This analysis involves undertaking the horizontal clustering location quotient (HCLQ). As opposed to the standard location quotient (LQ), the HCLQ measures the local importance of a sector given the overall importance of an area. Fingleton, Igliori and Moore (2005) develop this measure due to the lack of scale sensitivity of the LQ. The HCLQ measure corresponds to a spatial agglomeration of firms in a particular industry and the number of jobs in the industry that exceeds its expected number. The expected number is the number of jobs in one industry that would correspond to the area having the national share of that industry. Therefore, the HCLQ measures the difference between the actual and expected number of one particular industry. While the positive values of the HCLQ indicate spatially concentrated industries, the negative values indicate that the employment of a particular industry in a city is less than the expected employment and hence there is less concentration of that industry. The calculations of both measures are summarized in Figure 4.6.

Indicator	Formula
LQ	$LQ = (E_{ig}/E_{in})/(E_{og}/E_{on})$
HCLQ	$HCLQ = E_{ig}-\hat{E}_{ig}$

Description

E_{ig} is employment in sector *i* in region *g*

E_{og} is total employment in region *g*

E_{in} is national employment in sector *i*

E_{on} is national total employment

E_{ig} is actual employment in sector *i* in region *g*

$\hat{E}_{ig}$ is estimated employment in sector *i* in region *g* when LQ equals one

Figure 4.6 LQ and HCLQ calculations
Source: Ratanawaraha and Polenske (2007, 46).

Table 4.2 presents the LQs and HCLQs for each industry sector in Adana, Ankara, Bursa, Gaziantep, Hatay, Istanbul, Izmir and Kocaeli. Even though the two measures of geographical concentration provide similar results, the HCLQ has a stronger meaning for agglomeration or clustering. While the value of 1 and above indicates a concentration for a particular sector in the LQ, an arbitrary number of 1.5 has been adopted as an indicator for the HCLQ (see Tonts and Taylor 2010). The analysis suggests that each city has a degree of specialization. Of the eight cities, Istanbul has specialized in eleven corporations in Turkey, amongst which manufacturers of chemical products, not elsewhere classified has the highest degree of specialization across all the sectors in Turkey. This high degree of specialization was followed by Hatay with 6.05 even though this city is only specialized in one particular industry, for instance iron and steel basic industries.

Given the number of headquarters located in Istanbul, it is not surprising that Istanbul has the highest number of specialized corporations. Like Istanbul, Ankara has a number of intense agglomerations that are unique to that city. The most substantial of these sectors are mining and agglomeration of hard coal; manufacture of structural clay products; manufacture of machinery and equipment except electrical (nec); and production, collection and distribution of electricity. These results overlap with an earlier study that reveals engineering as a primary source of employment in Ankara (Akgungor 2006). Moreover, the analysis also resonates with newly developing industrial districts around Ankara in which Eraydin (2002) stresses stone based industries (particularly brick manufacturing) as an engine for growth in the area. She, for instance, emphasizes the importance of machinery production that is developed in the region (Eraydin 2002).

Bursa is another important player in corporate Turkey, with three substantial agglomerations: manufacture of textile goods except apparel; manufacture assembly of motor vehicles and repairing; and production, collection and distribution of electricity. Amongst these industries, Bosch has got a leading role for the manufacturing, assembling and repairing of motor vehicles. In terms of production, collection and distribution of electricity, BIS Energy and Zorlu Energy are very important energy providers in the city. Moreover, Bursa and its vicinity have been famous for textile production, characterized by accumulated knowledge in the traditional production culture. Production of textile in Bursa is in the form of flexible integration of large and small enterprises (Eraydin 1995).

Adana, Gaziantep, Izmir and Kocaeli have two specialized corporations in each. With manufacture of rubber products (nec), manufacture of food products (as the only significant sector across Turkey) are the two specialized industries in Adana. Similar to Bursa, Gaziantep has a significant concentration of manufacture of textile goods (except wearing apparel). With 80 per cent of employment in textile production, Gaziantep is an example of a less developed country industry district (Saral and Celebi 2002). Both Bursa and Gaziantep are the only cities specialized in that particular sector across Turkey. Gaziantep also has a degree of specialization in manufacture of rubber products (nec). Similar to Adana and Gaziantep, within the top 500 corporations, Turkey's third biggest city Izmir has a degree of specialization in the manufacture of rubber products (nec) as well. Though, this significance is dominated by tobacco manufactures. Japan Tobacco International and TTL Tobacco Industry are the two dominant tobacco

Table 4.2 Sectoral agglomeration of company headquarters

Sector	Location Quotient								Horizontal Cluster Location Quotient							
	Adana	Ankara	Bursa	G.antep	Hatay	Istanbul	Izmir	Kocaeli	Adana	Ankara	Bursa	G.antep	Hatay	Istanbul	Izmir	Kocaeli
Mining and agglomeration of hard coal	0.00	5.63	0.00	0.00	0.00	0.35	1.24	2.85	−0.29	4.93	−0.75	−0.26	−0.18	−3.75	0.19	1.30
Manufacture of food products	2.07	0.92	0.26	1.52	0.00	0.65	0.73	0.00	1.55	−0.41	−2.83	0.68	−0.92	−10.17	−1.09	−3.56
Tobacco manufactures	0.00	0.00	0.00	0.00	0.00	0.75	8.06	0.00	−0.13	−0.49	−0.35	−0.12	−0.08	−0.65	2.63	−0.32
Manufacture of textile goods, except wearing apparel	2.33	0.00	3.09	6.41	0.00	0.75	1.24	0.00	1.14	−3.20	4.74	4.22	−0.55	−4.24	0.58	−2.11
Manufacture of wearing apparel	0.00	0.00	0.00	0.00	0.00	1.70	2.02	0.00	−0.18	−0.66	−0.46	−0.16	−0.11	2.46	0.50	−0.43
Manufacture of wood and cork products	0.00	0.00	2.46	0.00	0.00	0.65	0.00	5.29	−0.15	−0.57	0.59	−0.14	−0.10	−1.09	−0.43	1.62
Manufacture of pulp, paper and paperboard	0.00	0.00	0.00	0.00	0.00	1.58	0.00	1.85	−0.22	−0.82	−0.58	−0.20	−0.14	2.58	−0.62	0.46
Printing, publishing and allied industries	0.00	1.52	0.00	0.00	0.00	1.70	2.02	0.00	−0.18	0.34	−0.46	−0.16	−0.11	2.46	0.50	−0.43
Manufacture of basic industrial chemicals, except fertilizers	0.00	0.00	0.00	0.00	0.00	1.41	2.02	2.31	−0.35	−1.31	−0.93	−0.32	−0.22	2.93	1.01	1.14
Manufacture of chemical products, not elsewhere classified (nec)	0.00	0.00	0.00	0.00	0.00	1.91	1.24	0.71	−0.57	−2.13	−1.51	−0.52	−0.36	10.51	0.39	−0.40
Petroleum refineries	0.00	0.00	0.00	0.00	0.00	2.04	0.00	1.85	−0.22	−0.82	−0.58	−0.20	−0.14	4.58	−0.62	0.46

Tire and tube industries	0.00	0.00	0.00	0.00	0.00	1.81	0.00	0.00	−0.11	−0.41	−0.29	−0.10	−0.07	1.79	−0.31	−0.27
Manufacture of rubber products, nec	8.52	0.00	0.00	6.25	0.00	0.85	3.02	0.00	2.65	−1.31	−0.93	1.68	−0.22	−1.07	2.01	−0.86
Manufacture of pottery, china and earthenware	0.00	0.00	0.00	0.00	0.00	1.81	3.23	0.00	−0.11	−0.41	−0.29	−0.10	−0.07	1.79	0.69	−0.27
Manufacture of glass and glass products	0.00	0.00	0.00	0.00	0.00	1.98	0.00	0.00	−0.18	−0.66	−0.46	−0.16	−0.11	3.46	−0.50	−0.43
Manufacture of structural clay products	1.47	2.36	1.11	1.61	0.00	0.58	1.04	0.60	0.32	3.46	0.20	0.38	−0.43	−5.70	0.08	−0.67
Iron and steel basic industries	0.67	0.90	0.51	0.00	7.35	1.03	0.71	1.63	−0.50	−0.58	−1.94	−1.36	6.05	0.94	−1.22	2.33
Non−ferrous metal basic industries	0.00	0.00	0.00	0.00	0.00	1.36	1.61	1.85	−0.22	−0.82	−0.58	−0.20	−0.14	1.58	0.38	0.46
Manufacture of machinery and equipment except electrical, nec	0.00	2.57	1.81	0.00	0.00	0.48	0.00	0.97	−0.42	2.44	0.90	−0.38	−0.27	−4.40	−1.18	−0.03
Manufacture of electrical industrial machinery and apparatus	0.00	1.26	0.59	0.00	0.00	1.17	0.56	1.28	−0.64	0.62	−0.68	−0.58	−0.41	2.18	−0.80	0.43
Manufacture assembly of motor vehicles and repairing	0.95	1.27	2.87	0.00	0.00	0.94	0.67	1.16	−0.06	1.06	5.22	−0.96	−0.67	−1.22	−0.98	0.41
Production, collection and distribution of electricity	0.00	2.81	3.98	0.00	0.00	0.87	0.00	2.85	−0.29	1.93	2.25	−0.26	−0.18	−0.75	−0.81	1.30

Note: LQs and HCLQs with the value over 1 are shaded.

Source: own calculations.

companies in Izmir. In comparison to earlier studies, the specializations of the top 500 enterprises are different from the general specialization of industries within Izmir as they suggest a heavy concentration of the manufacture of food products and production of machinery for Izmir (Kumral and Deger 2003). The top 500 corporations in Kocaeli, on the other hand, specialized in iron and steel basic industries and manufacture of wood and cork products. Kocaeli has got one of the highest GDP ratios per capita in Turkey. Elveren (2010), for instance, studies the wage distribution across Turkey and shows that Kocaeli's private sectors, such as metals, machines and equipment, leads to a staggering contribution to the area.

While it is difficult to specify the direct consequences of the specialization of large corporations in Turkey, some interpretations can be done. First is related to the privatization of government-owned large corporations. After privatization, a significant number of companies have been created. This leads to spillover effects contributing to conditions that foster the development of new firms, for instance sub-contractors. Second is the availability of human capital. Istanbul, Ankara, Izmir and Kocaeli have more connections with the global economy and to some extent they are the preferred locations for large corporations. Therefore, these cities have become attractive places for highly skilled and well paid workers. Third is related to the increasing international competition of the Turkish companies. Textiles and clothing, for instance, account for one-third of industrial employment and they are very competitive in international markets. Amongst the top 500 companies, more than 10 per cent of the firms specialize in textiles and clothing. Fourth is the development of second-tier cities. With the help of large corporations, not only do those cities play significant local and regional roles in Turkey, they also link Turkey and its regional economies into the global economy. Kocaeli, Bursa and Gaziantep are very good examples of this type of global connection due to their endogenous potentials.

Conclusion

This chapter examines the geography of the 'Top 500' large corporations in the Turkish urban system. It shows that not only the spatial organization of corporations remains important in urban systems but also they are linked to the dynamics of local and regional economies within developing countries. In terms of the location of companies, the spatial concentration of production profit, net value added and export figures, Istanbul dominates Turkey's urban system. Kocaeli, even though the distribution of headquarters in percentages is less than Ankara, outperforms all the other cities (including Ankara) in terms of production profit. Moreover, in terms of net value added, Ankara and Kocaeli remain the only cities in which the net output has the highest value amongst other indicators.

In terms of sectoral specialization, both Istanbul and Ankara have a number of intense agglomerations that are almost unique to these cities. Bursa remains another important player in corporate Turkey, with three substantial agglomerations. This is followed by Adana, Gaziantep, Izmir, Kocaeli and Hatay. An important finding from this study is that various cities have a distinctive internal structure comprising agglomerations of activity. While some of these activities are better performing, especially in the big metropolitan cities than others, the majority underperforms in average. More research is needed to explore the financial performances of these cities within the Turkish urban system. One example can be to analyse the impacts of privatization on the top 500 companies within each city. Also the analysis does not make a distinction between local and foreign-owned corporations at the moment. The specialization of these (local and foreign-owned) corporations, their share in manufacturing value added, involvement in low-medium-high technologies, R&D, and so on, might reveal the potential spill-over effects and impact on local economies.

Another issue is related to the dominant shareholders within the top 500 companies. The majority of these companies are owned or controlled by families. Understanding the social processes and values can lead to a flexible and dynamic corporate structure activity by facilitating new forms of establishment, knowledge and strategy. The significance of learning and innovation are crucial in urban studies. In Turkey, the geographical distribution of headquarters rely on the role of communication and transport hubs, proximity of business support services and availability of human and physical infrastructure. However, even though accessibility and economic development may seem complementary, when they are embedded in the political context, they do not always converge. Spatial patterns of accessibility seem fairly inert, in that only stringent spatial policies could change the overall accessibility in the long term. Therefore, the specialization of a particular area requires joint treatment by policy-makers. Changes in the patterns of control and specialization in Turkey are unlikely to happen rapidly in the future but it is still difficult to make a judgement within the global economy.

References

Akgungor, S. 2006. Geographic concentrations in Turkey's manufacturing industry: identifying regional highpoint clusters. *European Planning Studies*, 14(2), 169–97.

Amin, A. and Thrift, N. 1994. Living in the global, in *Globalization, Institutions, and Regional Development in Europe*, edited by A. Amin and N. Thrift. Oxford: Oxford University Press, 1–22.

Aytac, Z. and Sak, G. 2000. *Corporate Governance Assessment: The Turkish Equity Market*, World Bank. Available at: http://www.worldbank.org/ifa/rosc_cg_turkey.html.

Bachtler, J., Wishlade, F. and Yuill, D. 2001. Regional policy in Europe after enlargement, *Regional and Industrial Policy Research Paper*, 44, Glasgow: European Policies Research Centre, University of Strathclyde.

Bartlett, C. and Hedlund, G. (eds) 1996. *Managing the Global Firm.* London: International Thomson Business Press.

Bayirbag, M.K. 2010. Local entrepreneurialism and state rescaling in Turkey. *Urban Studies*, 47(2), 363–85.

Braczyk, H.J., Cooke, P. and Heidenreich, M. (eds) 1998. *Regional Innovation Systems: The Role of Governances in a Globalized World.* UCL Press, London.

Carroll, W.K. 2001. Westward Ho? The shifting geography of corporate power in Canada. *Journal of Canadian Studies*, 36, 118–42.

Carroll, W.K. 2007. Global cities in the global corporate network. *Environment and Planning A*, 39, 2297–323.

Celebioglu, F. and Dall'erba, S. 2009. Spatial disparities across the regions of Turkey: an exploratory spatial data analysis. *The Annals of Regional Sciences*, 45(2), 379–400.

Chan, S., Gau, G. and Wang, K. 1995. Stock market reaction to capital investment decisions: evidence from business relocations. *Journal of Financial and Quantitative Analysis*, 30, 81–100.

Conti, S. 1997. Global–local perspectives: a review of concepts and theoretical proposals, in *Interdependent and uneven development,* edited by M. Taylor and S. Conti. Aldershot: Ashgate, 15–56.

Cooke, P. 2002. Regional innovation systems: general findings and some new evidence from biotechnology clusters. *Journal of Technology Transfer*, 27(1), 133–45.

Cumbers, A., MacKinnon, D., and Chapman, K. 2003. Innovation, collaboration, and learning in regional clusters: a story of SMEs in the Aberdeen oil complex. *Environment and Planning A*, 35(9), 1689–706.

Demirag, I. *and* Serter, M. 2003. Ownership patterns and control in Turkish listed companies. *Corporate Governance*, 11, 40–51.

Dicken, P., and Malmberg, A. 2001. Firms in territories: a relational perspective. *Economic Geography*, 77 (4), 345–63.

Domański, B. 2004. Local and regional embeddedness of foreign industrial investors in Poland. *Prace Geograficzne Jagiellonian University*, 114, 37–54.

Domański, B. and Gwosdz, K. 2009. Toward a more embedded production system? Automotive supply networks and localized capabilities in Poland. *Growth and Change*, 40, 452–82.

Elveren, A.Y. 2010. Wage inequality in Turkish manufacturing by statistical regions, 1980–2001. *Review of Urban and Regional Development Studies*, 22(1), 55–72.

Elveren, A.Y. and Galbraith J.K. 2008. Pay inequality in Turkey in the neo-liberal era: 1980–2001. *University of Texas Inequality Project*, Working Paper 49.

Eraydin, A. 1995. Local development under the pressures of restructuring: the case of Bursa, Turkey, in *Recent Industrialization Experience of Turkey*, edited by B. Van der Knaap and R. Le Heron. New York: Greenwood, 155–76.

Eraydın, A. 2002. *Yeni Sanayi Odakları:Yerel Kalkınmanın Yeniden Kavramlastirilmasi*. Ankara: ODTU Mimarlık Fakultesi.

Eraydın, A. 2004. Bölgesel kalkınma kavram, kuram ve politikalarında yaşanan değişimler. *Kentsel Ekonomi Araştırmaları Sempozyumu 2003*, Bildiriler, 126–46.

Ersoy, A. 2012. *Dynamics and Drivers of Turkish Regional Development*, Birmingham: PhD Thesis, University of Birmingham.

Ersoy, A. and Taylor, M. 2012. Modeling local and regional economic development in Turkey: a curate's egg. *Growth and Change*, 43(4), 615–37.

Ertugal, E. 2005. *Europeanisation of Regional Policy and Regional Governance: The Case of Turkey. European Political Economy Review*, 3(1), 18–53.

Fingleton, B., Igliori, D. and Moore, B. 2005. Cluster dynamics: new evidence and projections for computing services in Great Britain. *Journal of Regional Science*, 45, 283–311

Florida, R. 2002. *The Rise of the Creative Class*. New York: Basic Books.

Gertler, M. 2001. Best practice? Geography, learning and the institutional limits to strong convergence. *Journal of Economic Geography*, 1, 5–26.

Gezici, F. and Hewings, G. 2004. Regional convergence and the economic performance of peripheral areas in Turkey. *Review of Urban and Regional Development Studies,* 16(2), 113–32.

Goodwin, W. 1965. The management centre in the United States. *The Geographical Review*, 55, 1–16.

Grabher, G. 1993. Rediscovering the social in the economics of interfirm relations, in *The Embedded Firm. On the Socioeconomics of Industrial Networks*, edited by G. Grabher. London: Routledge, 1–31.

Granovetter, M. 1973. The strength of weak ties. *American Journal of Sociology*, 78(6), 1360–80.

Granovetter, M. 1985. Economic action and social structure: the problem of embeddedness. *American Journal of Sociology*, 91, 481–510.

Guzik, R. and Micek, G. 2008. The impact of delocalisation on the European software industry, in *The Moving Frontier: The Changing Geography of Production in Labour-Intensive Industries*, edited by L. Labrianidis. Aldershot: Ashgate, 229–54.

Harrison, B. 1992. Industrial district: old wine in new bottle? *Regional Studies*, 26(5), 469–83.

Harzing, A. and Sorge, A. 2003. The relative impact of country-of-origin and universal contingencies on internationalization strategies and corporate control in multinational enterprises: world-wide and European perspectives. *Organization Studies*, 24, 187–214.

Heper, M. (ed.) (1991) *Strong State and Economic Interest Groups: The Post-1980 Turkish Experience*. New York: de Gruyter.

Hess, M. 2004. 'Spatial' relationships? Towards a reconceptualization of embeddedness. *Progress in Human Geography*, 28, 165–86.

Hirschman, A.O. 1958. *The Strategy of Economic Development.* New Haven, CT: Yale University Press.

Holloway, S. and Wheeler, J. 1991. Corporate headquarters relocation and changes in metropolitan corporate dominance, 1980–1987. *Economic Geography*, 67, 54–74.

Invest in Turkey. 2011. Republic of Turkey Prime Ministry Investment Support and Promotion Agency Website. Available at: http://www.dtm.gov.tr/dtmweb/indexen.cfm [accessed: 1 July 2011].

ISO. 2009. *Top 500 Enterprisers in Turkey*, Istanbul Chamber of Industry. Available at: http://www.iso.org.tr/en/index.aspx [accessed: 1 July 2011].

Kaymak, T. and Bektas, E. 2008. East meets west? Board characteristics in an emerging market: evidence from Turkish banks. *Corporate Governance: An International Review*, 16, 550–61.

Kirdar, M. and Saracoglu, S. 2007. Migration and regional convergence: an empirical investigation for Turkey. *MPRA Paper*, 2648. Available at: http://mpra.ub.uni-muenchen.de/2648/ [accessed: 1 July 2011].

Klier, T. 2006. Where the headquarters are: location patterns of large public companies, 1990– 2000. *Economic Development Quarterly*, 20, 117–28.

Klier, T. and Testa, W. 2002. Location trends of large company headquarters during the 90s. *Economic Perspectives*, 26, 12–26.

Kumral, N. and Deger, C. 2003. *An Industrial Cluster Study: As a Basis for the Aegean Region's Development Policy*. Paper presented at the Regional Studies Association International Conference, 12–15 April 2003.

MacKinnon, D., Cumbers, A., and Chapman, K. 2002. Learning innovation and regional development: a critical appraisal of recent debates. *Progress in Human Geography*, 26(3), 293–312.

Martin, R. and Sunley, P. 2001. Rethinking the 'economic' in economic geography: broadening our vision or losing our focus? *Antipode*, 33, 148–61.

Myrdal, G. 1957. *Economic Theory and Under-developed Regions*. London: Duckworth.

Oinas, P. 1997. On the socio-spatial embeddedness of business firms. *Erdkunde* 51, 23–32.

Onis, Z. 2011. Power, interest and coalitions: the political economy of mass privatization in Turkey. *Third World Quarterly*, 32(4), 707–24.

Pavlínek, P. 2004. Regional development implications of foreign direct investment in Central Europe. *European Urban and Regional Studies*, 11, 47–70.

Perroux, F. 1955. Note sur la notion de 'pole de croissance' [A note on the notion of growth poles]. in D.L. McKee, R.D., Dean, W.H. Leahy (eds). 1970. *Regional Economics*, 93–104, New York, Free Press..

Perroux, F. 1973. Multinational investment and the analysis of development and integrated pole. *Economies et Societies* 24, 831–68.

Phelps, N.A. 1993. Branch plants and the evolving spatial division of labour: a study of material linkage in the northern region of England. *Regional Studies* 27(2), 87–101.

Phelps, N.A.; MacKinnon, D.; Stone, I.; and Braidford, P. 2003. Embedding the multinationals? Institutions and the development of overseas manufacturing affiliates in Wales and North East England. *Regional Studies*, 37, 27–40.

Polanyi, K. 1944. *The Great Transformation*. New York: Farrar & Rinehart.

Porter, M.E. 1998. *On Competition*. Boston MA: Harvard Business School Press.

Quark, A. 2007. From global cities to the lands' end: the relocation of corporate headquarters and the new company towns of rural America. *Qualitative Sociology*, 30, 21–40.

Ratanawaraha, A. and Polenske, K.R. (2007) Measuring the geography of innovation: a literature review, in *The Economic Geography of Innovation*, edited by K.R. Polenske. Cambridge: Cambridge University Press, 30–59.

Rodrik, D. 2010. *Structural Transformation and Economic Development*. Policy Research Foundation of Turkey (TEPAV). Ankara.

Sabel, Ch. F. and Zeitlin J. 1997. Stories, strategies, structures: rethinking historical alternatives to mass production, in *World of Possibilities. Flexibility and Mass Production in Western Industrialisation,* edited by C.F. Sabel and J. Zeitlin. Cambridge: Cambridge University Press, 1–33.

Sahin, S.B. 2010. Privatization as a Hegemonic Process in Turkey. *Journal of Contemporary European Studies*, 18(4), 483–98.

Saral, G. ans Celebi, D. 2002. *The Innovative Capacity and Learning Capability of Turkish Regions.* Paper presented at ERC/METU International Conference in Economics VI, Ankara, 11–14 September 2002.

Sassen, S. 2006. *Cities in a World Economy* (3rd edn), Thousand Oaks, CA: Pine Forge Press.

Scott, J. 1997. *Corporate Business and Capitalist Classes*. New York: Oxford University Press.

Storper, M. 1997. *The Regional World: Territorial Development in a Global Economy*. New York: Guilford Press.

Taylor, P. 2004. *World City Network: A Global Urban Analysis*. London: Routledge.

Taylor, M. 2005. Embedded local growth: a theory taken too far?, in *Learning from Clusters: A Critical Assessment from an Economic–Geographical Perspective*, edited by R. Boschma and R. Kloosterman. Dordrecht: Springer, 69–88.

Taylor, M., and Ersoy, A. 2011. *Understanding Local Industrial Growth: From Theories to Pragmatic Local Policies.* Policy Research Foundation of Turkey (TEPAV), Ankara.

Taylor, M. and Thrift, N. 1980. Large corporations and concentrations of capital in Australia: a geographical analysis. *Economic Geography* 56, 261–80.

Taylor, M. and Thrift, N. 1982. Industrial linkage and the segmented economy, 1. *Environment and Planning A,* 14, 1601–13.

Taylor, P. 2007. *Cities in Globalisation: Practices, Policies and Theories*. London: Routledge.

Taylor, P., Ni, P. and Derudder, B. 2009. The way we were: command and control centres in the global space-economy on the eve of the 2008 geo-economic transition. *Environment and Planning A*, 41, 7–12.

Tekeli, İ. 2004. *Bir Bölge Plancısının* Çözmeye Çalıştığı *Sorunun Büyüklüğü Karşısında Yaşadığı* İç *Huzursuzlukları*. DPT Kentsel ve Ekonomik Araştırmalar Sempozyumu, Cilt II: 244–62.

Tonts, M. and Taylor, M. 2010. Corporate location, concentration and performance: large company headquarters in the Australian urban system. *Urban Studies*, 47(12), 2641–64.

van Egeraat, C., and Jacobson, D. 2005. Geography of production linkages in the Irish and Scottish microcomputer industry: The role of logistics. *Economic Geography*, 81, 283–304.

Welch, F. 1970. Education in production. *Journal of Political Economy*, 78(1), 35–59.

Yeung, H., Poon, J. and Perry, M. 2001. Towards a regional strategy: the role of regional headquarters of foreign firms in Singapore. *Urban Studies*, 38, 157–83.

Yeung, Y. and Li, X. 2000. Transnational corporations and local embeddedness: company case studies from Shanghai, China. *The Professional Geographer*, 52(4), 624–35.

Chapter 5
Foreign Direct Investment in Estonia—Understanding the Impact of Public Policies on Local Embeddedness and Networking in the Food Retail and Related Industries

Egert Juuse, Sylvi Birgit Endresen, Rainer Kattel

Introduction

Many scholarly works have been published on foreign direct investments (FDI) in developing and transition economies with the focus on inward FDI's impact on the host economy. Even though modern retail in these economies has been extensively studied, analytical research on FDI in the food retail industry in Estonia represents a missing link. On the theoretical level, Estonia is a significant example. Being one of the more successful transition economies, Estonia's economic policy framework is perhaps the clearest example of applying a neoliberal policy toolbox in a transition context and beyond (see Thorhallsson and Kattel 2012).[1] Hence, the study of Estonian retail offers an opportunity to understand the impact of neoliberal policies on retail and related industries. As the neoliberal policy toolbox would prescribe, Estonia has not used any foreign investment management policies beyond macroeconomic reforms oriented towards price stability and low taxes. Consequently, while being a small and extremely open economy, the Estonian experience offers relatively unique opportunity to understand the consequences and impacts of an unmanaged FDI policy. On the international comparative basis, the Estonian case reveals the challenges for the host economy that have resulted from rapid developments during a relatively short period (less than 15 years), compared to the decades-long transformation of retail business in the Western hemisphere.[2]

We analyze the role of FDI in the Estonian retail industry in forming the relations between retail and related industries, that is, the embeddedness of FDI in retail in the Estonian economy and the resilience of domestic players to outer

1 For a useful discussion of varieties of capitalisms emerging from the transition context, see Drahokoupil (2007, 2008).

2 See Wrigley and Lowe (2002) on the historical development of modern retail.

pressures. The first section presents the theoretical framework for analyzing the embeddedness of and linkages between Estonian retail and related industries by outlining stylized facts, while the next section deals with the analysis of the Estonian case, followed by the discussion section and concluding remarks. The empirical analysis relies on statistical information and secondary data, based on previous surveys and journalistic literature, as well as interviews conducted with key stakeholders in 2010–11: the Estonian Chamber of Commerce and Industry, the Estonian Traders Association, the Association of Estonian Food Industry, the Estonian Chamber of Agriculture and Commerce, the Estonian Dairy Association, the Estonian Ministry of Economic Affairs and Communications, and the Estonian Ministry of Agriculture.

Theoretical Considerations—Analytical Framework and Stylized Facts

The current analysis is based on the foundations of the heterodox economics tradition, which views the capitalist economic system as cyclical and institution-rich (Lavoie 2009, Keen 2011). Pursuant to this tradition, the study embraces a holistic approach by focusing on economic and political institutions as well as the power relations between economic sectors that determine the level of economic activity. For understanding the current state of the economy, the changes in the productive structure with the notions of path-dependency (irreversibility) and lock-in effects need to be addressed.

From the heterodox perspective, modern capitalist economies are seen as monetary production systems with firms operating in an imperfectly competitive environment, that is, oligopolistic markets with interdependent business relations. Firms endeavor to increase their size and market share by utilizing the mechanisms of economies of scale and scope. Efficiency gains achieved by cost advantages and increasing returns enable large businesses to out-compete small enterprises and thus render perfectly competitive markets into an oligopolistic situation. Economies of scale are an important reason why most industries are dominated by a small number of large companies. Yet, by acknowledging the importance of systemic monetary structure with interlinked sectoral balance sheets and cash flows, balanced as well as stable demand–supply relations between different sectors, that is backward and forward linkages, need to be established for sustainable growth. Therefore, heterodox tradition calls for state regulations and economic policies as institutional constraints for stabilizing the inherently unstable and cyclical market economy with deepening sectoral imbalances (see Minsky 2008, Lavoie 2009, Keen, 2011). In essence, such a framework assumes that government intervention is present in a given economy, notwithstanding the prevailing ideologies and power coalitions; that is, a lack of policy is also a policy with discernible impact on industry dynamics.

Hence, for analyzing the Estonian food industry, a network perspective (Markusen 1999a, Wrigley and Lowe 2002, Coe and Wrigley 2007, Dicken

2011) is applied to gain insight regarding the transformation of retail and related sectors affected by transnational companies (TNC). This type of embeddedness in Hess's (2004: 180) typology is addressed in the study of inter-firm relations with local suppliers and other extra-firm relations within a specific institutional and regulatory framework (see Figure 5.1). Network embeddedness can be studied regardless of the location of actors; however, aspects of territorial embeddedness are also covered, which occurs "when foreign actors build considerable links to the actors present within the respective host localities" (Hess, 2004). Furthermore,

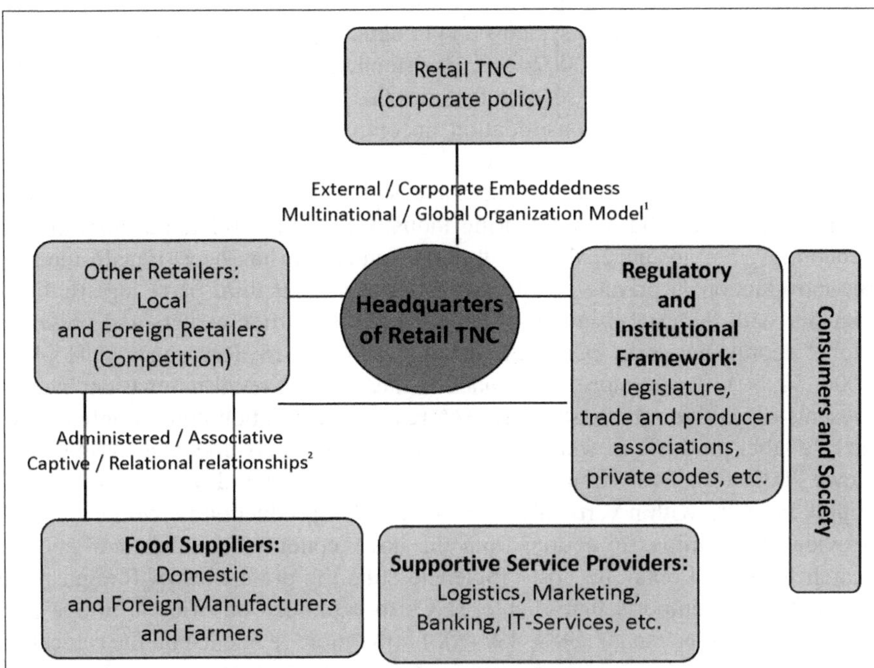

Figure 5.1 Retail network and societal embeddedness perspective

Note: [1] The *multinational (external) organization model* is based on decentralization, a high degree of autonomy, informal coordination, simple financial control, and local focus. By contrast, the *global (corporate) model* signifies stringent control, impelling coordination by headquarters, centralization, and standard approach (Dicken 2011: 121–41). [2] *Administered (close-control or dependent) relationships* imply supply relationships in which one party coordinates activity on the basis of power (short-term contracts, inequality in trading). Retailers assume control of and coordinate many of the supply chain functions, whereas *associative (arms-length or developmental) relationships* involve a high level of collaboration and cooperation between retailers and manufacturers—value-adding partnerships, often long-term, and centered on information sharing networks (Wrigley and Lowe 2002: 55, Dicken 2011: 438).

Source: own elaboration.

the Estonian political, economic, and social context is captured by the concept of societal embeddedness, that is, the exposure of retail TNCs to the "different cultures of their foreign network partners" (Hess 2004). In addition to local embeddedness in Estonian society, TNCs import their "cultures" and may seek to transform the host society according to their priorities, as will be revealed in this chapter.

In line with the notion of societal embeddedness, the key to the successful internationalization of retail lies in the adaptation to local circumstances. Nonetheless, the primary way for retail TNCs to reap profits has been to transfer the operational techniques and business models to host markets in order to gain control over the production systems and to obtain bargaining power (see Dawson 2007: 390–94, Humphrey 2007: 449, Christopherson 2007: 464). Studies on FDI in food retail have found several features that are evident in host economies. Firstly, there are signs of consolidation, accompanied by vertical integration, and secondly, there is the introduction of higher value added products, coupled with private label products. Both aspects have had a significant impact on business operations in the food manufacturing industry and other related sectors in host economies. For instance, the institutional landscape has been transformed by the introduction of private quality standards as well as third-party logistic firms together with the establishment of centralized distribution centers and preferred global supply networks (Coe and Wrigley 2007: 9–16, Reardon and Hopkins 2006: 523–9). Furthermore, the academic literature on retail internationalization presents a positive relationship between retail concentration and the intensity of private label products as well as private label supply networks (see Wrigley and Lowe 2002: 64). Even though the latter may involve the development of local supply linkages within vertically integrated global production networks and thus provide opportunities to source from the host economy for other markets, the search for operational and cost efficiencies tilts the preference of foreign retail chains towards imports that enable TNCs to capitalize on price arbitrage (see Dicken 2011: 436, Jacobs 1984: 195–6). Furthermore, evidence on the relocation of suppliers and other partner companies from home to host economies could be considered as covert imports (see Reardon, Henson and Berdegue 2007: 425–6), leading to an enclavization[3] processes within the host economy (see Gallagher and Zarsky (2007) on the concept of an enclave economy).

Irrespective of the host economy or cross-border dimensionality, retail chains tend to source from a few large suppliers for each product category due to the requirements imposed on suppliers in terms of product quality, safety, quantity, delivery consistency, and so on that are difficult for small producers to comply with. This has indicates the need for food manufacturers and farmers to adapt

3 One could consider FDI in retail as a *satellite platform* (see Markusen 1999b: 21–41) that reinforces the reliance on finances, technical expertise, and services external to the region. Such a high dependency on foreign capital implies hazardous lock-in effects and hence undermines the host economy's autonomy and economic sustainability.

to the demands of the retail sector (Humphrey 2007: 433–47, Reardon, Henson and Berdegue 2007: 406–26, Weatherspoon, Reardon 2003: 333–4, Blythman 2005: 184–92). In these circumstances, the development of modern retail has entailed positive feedback mechanisms, that is, the self-reinforcing dominance of interlinked large companies (Simms 2007: 6–13). As a result, retail chains tend to stock national brand leaders plus private labels, which leads to the demise of small producers. All in all, the emerging imbalances between retail and related industries, incurred by the increased concentration and control over a range of functions on the part of retailers, have made it possible to capitalize on price arbitrage and to demand price discounts from suppliers with the threat of delisting (see Wrigley and Lowe 2002: 51–3). With requirements that incur extra expenditure, suppliers have been unable to gain from economies of scale and often sell the produce under the cost of production. These tendencies have made producers susceptible to financial losses that could culminate in bankruptcies as well as increased unemployment (see Blythman 2005: 171–83, 258–78). Moreover, established inequalities in contractual relationships between retailers and suppliers have enabled retail chains to apply exploitative practices such as fees and fines for various forms of misconduct on the suppliers' side. Such patterns indicate an *oligopsonistic* condition (see Aoyama 2007: 472, Blythman 2005: 3–15, 148–92), whereby risks and costs are passed down the value chain.

As monopolistic or oligopolistic markets entail intrinsic vulnerabilities as well as systemic risks for the whole economy (see Lavoie 2009: 12), the possibilities of alleviating this fragility depend on the extent of government intervention in the economy. The potential for government intervention in the retail industry directly via competition and trade regulations or indirectly with taxes cannot be underestimated when broader macroeconomic as well as industrial policy measures are expected to have positive effects in stimulating economic growth and supporting capacity building in manufacturing industries. For these reasons, countries with a long tradition of modern retail, for example, the UK, the USA and Scandinavian countries, have relied on alternative interventionist policies[4] to direct developments in food retail and related industries in their economies.

4 Examples are restrictions on the ownership and *anti-market-extension* mergers; regulations on employment and "below-cost" selling; pricing and payment regulations to clamp down on unfair price war practices; requirements to explain the value of TNC to the host economy in terms of increasing employment, infrastructure investments, skills, and exports (see Coe and Wrigley 2007: 24–5, Reardon and Hopkins 2006: 532–6, Wrigley and Lowe 2002: 119–20, Blythman 2005: 323–4, Simms 2007: 53–4). Concerns over the increasing imports and balance of payments problems can be mitigated by the restrictions on profit repatriation as well as requirements to source from local producers (see Weatherspoon and Reardon 2007: 347, Dicken 2011: 286).

The Formation of the Estonian Food Retail Industry

After the collapse of the Soviet Union in 1991, Estonia faced the challenge of instituting a capitalist free market economy with a supportive institutional-legal framework. Transformations in the economy did not leave the retail and related sectors untouched, where the subsidiaries of retail TNCs were the catalysts of major adjustments by setting the path and creating spillover effects on the local market. During the last 15 years, Estonia has witnessed aggressive expansion of both domestic and foreign retail chains of diverse formats, geographic dispersion patterns, and market-targeting strategies. As a result of takeovers as well as organic expansion of retail chains, the food retail landscape has changed drastically over the years by gradually crowding out independent small shops and open markets in the fierce cost-based competition. While less than 40 percent of the retail market was controlled by the chains and around 15 percent by foreign owners in 2002, the same figures for 2010 were 78 percent and 45 percent, respectively (Centre of Registers and Information Systems 2012, see also Chapter 6 of this book).

Today, one can observe *pan-Balticanization* of the retail market in terms of the same foreign companies operating in all three Baltic States (Kaljuvee 2007), as the expansion of Lithuanian, Swedish, and Finnish retail chains in Estonia and other Baltic countries can be explained by the cultural proximity of investor and host countries. Big chains like Tesco and Carrefour have not invested yet, probably due to the relatively small markets and linguistic–cultural differences (Merilai 2011).

The interweaving of different business cultures in a competitive environment has gradually led to the standardization of service provision and business relations in the Estonian retail landscape in line with the best practices of Nordic countries. Highlighting packages deposit and recycling systems, the Estonian Traders Association underlines that in general, the foreign presence is positive: "The main impact of foreign retailers on Estonian economy has been the transfer of business culture, management best practices, and equipment as well as capital inflows" (Interview, M. Merilai 2011).

Aside from the aforementioned advantages, the increasing dominance of foreign ownership in the food retail industry has entailed several threats and risks for food retail and related industries. As competition among food retailers has intensified, pressures have been transmitted down the value chain to the producers of food products. This is reflected in the better bargaining position of chains against producers and vendors as compared to independent stores, which enables the chains to lower the prices (Taube 2005). On the other hand, there is a clear recognition in the Estonian Traders Association that when companies are run from abroad, this may limit local linkage formation. A case in point is Rimi Baltic which is managed from Latvia, limiting the autonomy of local managers (Merilai 2011).

Even though it has been found that Estonian consumers tend to be loyal to or at least have preference towards local products, which are considered to be safer and more reliable (Potisepp 2010), in deteriorating economic conditions people have shifted their purchases towards imports, which are cheaper (EIER 2010).

The Association of Estonian Food Industry emphasizes the Estonian consumers' preference for local products and their consciousness regarding origin, but claim that the price factor is important and therefore imported food may invade the market (Potisepp 2011). Thereby, during the economic recession, the ensuing effects of price competition on food retailers can lead to the deterioration of the economic sustainability of local food producers which cannot compete with cheaper imports: "Local producers are restricted in dictating the developments and affecting prices, as retailers can turn to imported goods" (Interview, E. Sokk, Estonian Ministry of Agriculture 2011).

Therefore, the main channels whereby FDI in retail has influenced the Estonian food industry are: 1) imports by relying on global production networks, 2) local procurement, and 3) private labels that reflect overt or covert aspirations towards vertical integration and are manifest in both foreign trade as well as cross-border and local linkages. The following sections address these issues and, in particular, how these patterns have emerged and what the enabling factors have been.

A Fragile Situation for Estonian Food Manufacturers

Aside from local competition, Estonian food manufacturers face additional pressures from imported goods, although openness to international trade and its high share in GDP (reaching 200 percent in Estonia in 2000) is not an uncommon feature of small economies. Yet, the high exposure to foreign competition and its detrimental effect on the local economy has been aggravated by the weaknesses in the regulation of production and trade. Unlike their European competitors, the Estonian farmers have not been equally subsidized, implying high imports (Sõrmus 2011).

Moreover, the Estonian economy has suffered from "Dutch disease" syndrome in terms of an appreciating real exchange rate due to the influx of foreign capital and accompanying increases in domestic prices, which have had a deleterious effect upon exports, but subsidized consumers by making imports cheaper. In these circumstances, surges of food imports that have negatively affected local producers by decreasing their sales and profits have not been rare (Erilaid 2004).

Furthermore, Estonian producers cannot often supply the volumes preferred by the big chains (Palts 2011). Here, the risk for local producers of being crowded out by imports has been amplified by foreign food retailers' reliance on corporate business model, whereby the decisions regarding the assortment of produce are made abroad with the preference towards foreign suppliers (EIER 2010, Potisepp 2011). This is illustrated by the cases of *Rimi Eesti Food Ltd* and *Maxima Ltd* (with the market shares of 21 percent and 15 percent, respectively, as of 2010): decisions about the assortment of produce and other business related issues are made in Latvia and Lithuania, respectively. By using the same retailing methods as applied in Lithuania and Latvia, the strategy of *Maxima Ltd.* has been to centralize procurement in order to squeeze out the lowest possible prices to the consumer, implying the dominance of cheaper foreign products. Local Estonian producers have not endeavored to market their goods through *Maxima Ltd* due to their higher price category products and unfavorable contractual terms (Ronk 2004, EPLO 2004, Reimer 2009). In contrast, by operating a purchasing department in Estonia, Finnish *Prisma Peremarket Ltd* (with the market share of 7 percent in 2010) has procured a large share of products from local suppliers in order to meet the demand of various local consumer segments, as has the local chain *Selver Ltd* (with the market share of 19 percent in 2010) (Kallas 2009, Lohk 2008, 2009).

Nevertheless, local producers have expressed concerns about the dominance of foreign retailers and their preference to procure from foreign producers, operating either in Estonia or abroad (Neudorf 2002). A case in point is *Prisma Peremarket Ltd* who decided to cooperate with Finnish *Fazer Amica* to supply the retail chain with processed food products such as frozen ready meals (Villak 2008). Due to foreign retailers' outward oriented supplier strategy, the challenge of enclavization and lack of backward linkages in the Estonian economy is acknowledged by the Estonian Traders Association. For instance, *Prisma Peremarket Ltd* benefits from their contact with partners abroad, even regarding IT support services (Merilai 2011).

Tensions between Producers and Retailers

The lack of resilience of domestic producers to withstand the pressures from the retailers could be attributed to the pyramidal structure of the value chain. Views differ between the retail business community and the producers on just how asymmetric power relations are. An informant in the Ministry of Economic Affairs and Communications holds this opinion: "It is evident that in a pyramid scheme, where there are six retail chains, around 40 dairy producers and hundreds of farms, leverage tends to be in favor for retailers" (Interview, M, Sõrm 2011). A spokesman for the Estonian Chamber of Agriculture and Commerce concurs: "Clearly, retailers dictate what products are to be produced. The quality requirements need to be guaranteed by producers" (Interview, R. Sõrmus 2011).

The Estonian Traders Association also perceives the conflict of interest between the parties but considers disagreement regarding the distribution of profits in the value chain as part of a normal business life: "The confrontation between producers and retailers is overemphasized and overplayed in the media" (Interview, M. Merilai 2011).

Her views are shared in the Estonian Chamber of Commerce and Industry with some concessions. Denying that retailers generally dominate producers, it is admitted that recently, changes have occurred: "The attitude of foreign retailers has been somehow different and there have been issues concerning the enforcement of contractual terms upon producers" (Interview, M. Palts 2011).

The "issues" referred to may be termed *predatory business practices* that reflect asymmetrical relations. In addition to requiring discounts and bonus payments from suppliers, there have been cases where retailers have imposed various penalties on producers as well as charges related to logistics, administration, and marketing. In general, there tend to be deductions from producers' sales price of up to 25 percent to cover the costs related to such services (Bank 2010, Saron 2011). The Estonian Chamber of Agriculture and Commerce refers to general exploitative practices such as payment delays and extra charges, driven by intense competition between retailers seeking a dominant market position (Sõrmus, 2011). Such practices are detrimental to suppliers in farming as well as manufacturing. In the Association of Estonian Food Industry, contracts between the parties are seen to favor retailers. Penalties and delisting are considered part of their corporate strategies. The major barrier for producers is a lack of alternative marketing channels (Potisepp 2011). For instance, rigidness in negotiations, threats of delisting, and penalizing contractual terms with suppliers have been the features of *Maxima Ltd.*, whereas any revolt by the suppliers has been suppressed by the argument of a leading marketing channel (Karotamm 2008).

Hence, in order to withstand the pressures from retailers, imports, and foreign companies, local food manufacturers have responded with consolidation and market concentration into the hands of a few companies operating in different product categories. On average, the five largest companies in all food categories have gained control of around 60–75 percent of the market (Lindpere et al. 2011). Concentration is evident to the Estonian Chamber of Agriculture and Commerce: "Companies are getting larger, which is one way of resisting the dominance of retail. In fresh vegetable market, there are few players, less than ten" (Interview, R. Sõrmus 2011).

He has found that the marketing situation improved as farmers began to form cooperatives over the last five years. A similar proposition has been made by the Estonian Dairy Association:

> Today, there are too many dairy producers for a limited local market. Most of the companies in the food industry are small or medium-sized enterprises, which has given them more flexibility in rearranging product portfolios. Yet, larger foreign companies tend to have an advantage of economies of scale and a larger market share. Thus, more concentration is needed, but the question is to what point. (Interview, T. Saron 2011)

An alternative to increased concentration has been the establishment of network systems. In the meat sector, both local and foreign manufacturers have built up their own local supply systems, but have also depended on imported packaged *ready-to-be-sold* products or raw materials to be processed into final products (Sõrmus 2011, Merilai 2011). Networking is reflected in a higher specialization, as some of the product lines have been outsourced to competitors to optimize the costs (Sokk 2011).

The concentration of capital can also be observed in farming. In contrast to other Central and Eastern European countries (CEE) countries, except for the Czech Republic, large farms dominate in Estonia (Saron 2011, Sokk 2011), and according to the Estonian Chamber of Agriculture and Commerce, they will become larger and the number of firms will decrease (Sõrmus 2011). The average area per agricultural holding (farm) was 47.7 hectares in Estonia compared to 14.1 hectares in the EU27, 21.5 and 13.7 hectares in Latvia and Lithuania respectively. At the same time, the total number of agricultural holdings in Estonia is the third smallest in the European Union—19,700 compared to 83,000 in Latvia and 200,000 in Lithuania (Eurostat 2011).

Overall, these consolidation trends have been strengthened by retail chain policies. By the mid-2000s, all retail chains had narrowed the number of local suppliers in the meat and dairy categories (Männiste 2006). The focus of food retailers has been on procuring from large manufacturers, who in turn have shifted their marketing rhetoric, arguing that the procured amounts by small retailers are insignificant and thus running business with them is inconvenient or even risky (PM Online 2002, Johanson 2007, Raudoja 2007). As a consequence, the possibilities for small producers to supply retail chains have deteriorated due to inconsistent quantities and deliveries (Niitra 2009, Ninn 2009). The Estonian Traders Association observes a dualism in retail outlets: supermarkets for large manufacturers and open markets as well as small shops for the small producers. This is attributed to a lack of cooperation between small producers (Merilai 2011). Cooperation as a solution to the problem of scale is recognized in the Estonian Chamber of Agriculture and Commerce as well. The major barriers are considered to be quality assurance and supply continuity (Sõrmus 2011). Even though all stakeholders have emphasized the importance of cooperation, extensive vertical integration among the layers of the value chain in the food industry up to the retail segment has not occurred (Pällin 2007). Vertical integration has been mainly observed where foreign companies have bought farms and invested in agribusinesses. However, the Estonian Chamber of Commerce considers conservatism among consumers regarding mass production to be hampering the process (Sõrmus 2011).

Introduction of Private Label Products and the Transformation of Wholesale

Retailers' own brand product—a private label—is a phenomenon that is directly attributable to FDI in Estonian retail (Sõrmus 2011). This may be seen as a covert attempt by retail chains to deepen their vertical integration or firmly root the presence of imported goods. Vertical integration that increases the value capture at the retail end of the value chain could be seen as the rationale behind the strategy of private labels. The informant in the Ministry of Economic Affairs and Communications recognizes this, but weighs it against the price advantages for the consumer (Sõrm 2011).

With regard to private label products and the consequences of this innovation, views among stakeholders tend to differ. Even though these low price products remove the marketing costs for producers, meet the needs of price-sensitive consumers, and enable producers to absorb excess capacities, private labels create confusion for consumers by losing distinctive features and becoming anonymous. From the producers' perspective, private labels cause additional competition by forcing them to lower production costs or by being imported due to the inability of Estonian food manufacturers to supply retailers with the required quantities (see Kallas 2011, Sokk 2011, Sõrmus 2011). Critics claim that private labels are designed to increase market share, gain control over product development and marketing, as well as mimic best-selling items. In this way, the possibilities to compensate research and development (R&D) expenditures of innovators are undermined and local producers' own brands are eventually outcompeted by the lower prices of private labels (Pällin 2007, Potisepp 2011, Sõrmus 2011). In particular, product categories that have high R&D costs as well as producers of *organic food* are sensitive to this (Saron 2011).

The Estonian Chamber of Agriculture and Commerce, on the other hand, highlights the new export opportunities that this innovation opens up for product segments that operate with excess capacities. The quality of Estonian produce is still questioned outside the volatile Russian market and private labels thus represent an opportunity (Sõrmus 2011). Hence the international networks provided by foreign retail chains might be conceived of as an export platform for Estonian local producers. For instance, *Prisma Peremarket Ltd* has provided the opportunity to market local food to other Baltic States and Finland under the chain's private label (Reimer 2010). Even though several obstacles have occurred in exporting, for instance, insufficient financial resources for promotion, low competitiveness, and low productive capacity, the quantities of exported products, either under producers own or a retailer's brand, have seen an upward trend (Pällin 2007, Potisepp 2010).

Aside from close-control relationships with suppliers, even deeper vertical integration has taken place by gaining control over different functions, for example, retailers' diversifying into deli, pastry, and bakery businesses by establishing wholly owned production units. In 2006, *Rimi Eesti Food Ltd* opened its central kitchen that services all its stores with ready-made meals in 13 different product categories with over 300 products (Kirsipuu 2006). Furthermore, the logistics and wholesale landscape has been transformed by retail chains. Traditionally, companies in the Estonian wholesale market tended to operate as sales representatives or distributors of imported goods. The early 2000s, however, witnessed the arrival of logistics centers which specialized in servicing large retail chains and paved the way for the concentration in the wholesale market (Merilai 2011). By 2010, all major retail chains were operating their own centralized distribution centers that enabled them to achieve efficiency gains on the distribution side and firmer control over the supply chain. Although suppliers were supposed to benefit from the increased capacity of freight transportation and less administration, they have been burdened with logistics fees imposed by retail chains. In addition, producers have been required to make substantial investments into information technology systems to comply with the operation systems of distribution centers (Raig 2011). Thus, out of the available channels of influence (see Munson, Rosenblatt and Rosenblatt 1999), retail chains in Estonia have mainly used pricing control, performance requirements, and control over the supply chain structure.

Governments' Positions—Belief in the Non-Interventionist Free Market

The finding of the study (see Potisepp 2006) that more than 80 percent of Estonian consumers prefer local food has been reflected to some extent in recent government initiatives to support local producers. In 2010, the government initiated a measure to fund the groupings of independent local farmers that would enhance their business performance, cooperation, and improve their bargaining position against manufacturers and retailers (Maasalu 2010). In particular, promoting so-called alternative food networks is high on the agenda (see Tregear 2011): both the Estonian government and the EU have stressed the importance of enhancing direct marketing among producers, and a priority of the Ministry of Agriculture has been to foster the consumption of short-distance food produce. In addition, cooperation between the food industry and R&D institutions has been supported in order to develop higher value-added products (Sokk 2011). The Estonian Dairy Association would welcome the support to small producers' marketing and limits to the size of shops to curb the retail chain power. However, these interventionist measures are met with skepticism, as Estonia would not be able to run such policies alone (Saron 2011).

Given the importance of the EU funds, major resentments have been related to the discrimination of Estonian agribusinesses within the Common Agricultural Policy (CAP) in terms of the differential allocation of agricultural subsidies between the new and the old members of the EU (Sokk 2011, Sõrmus 2011). Further concerns have been raised over the lack of public investment in the food manufacturing sector as within the CAP, the main focus of the government support has been on small farms (Merilai 2011, Saron 2011).

All things considered, public policies that have targeted productive sectors have provided the leeway to alleviate the imbalances in food retail and related businesses, as no specific legal acts have been enforced to deal with retailers directly due to the constraints that stem from international treaties and agreements, for example, within the WTO or the EU. It is producers who are mostly affected by regulations, not retailers (Palts 2011).

There is a strong belief in a free market economy among the stakeholders, very clearly expressed by M. Merilai of the Estonian Traders Association: "It is the market that self-regulates and allocates resources most efficiently" (Interview 2011).

The policy of the Estonian Food Producers' Association is equally free market friendly, as "government should not intervene into everyday business or do it as little as possible." (Potisepp 2011). Hence, they would not support any restrictions; all players should be treated equally and supporting measures that distort competition should be avoided. That interventions in markets for food may be necessary for socioeconomic reasons is, however, recognized in the Estonian Ministry of Agriculture (Sokk 2011). The Estonian Chamber of Agriculture and Commerce finds that the economy should be free of government intervention, unless competition law is infringed. But the concentration of capital, leading to retail giants, may pose problems; therefore, restrictions could be imposed during recessions "to rein the dominance of retailers and enforce stricter competition law by lowering the 40 percent threshold to eliminate the exploitative practices" (Sõrmus 2011).

In accordance with these standpoints, government coalitions have adhered to neoliberal policies that have obliterated any consideration of strong interventions such as the ceilings on profit margins in food produce or the requirements for retailers to procure from local producers, although similar restrictions have been contemplated in Lithuania, Latvia, and also in Slovakia (Äripäev 2009, Merilai 2011). To make things worse, the affinity of governments towards conservative fiscal policy reveals its pro-cyclical nature, as the tax base was lowered during the years of rapid economic expansion, while several taxes were introduced or increased during the post-crisis years (value added tax increased from 18 to 20 percent, plastic bag and sales tax were introduced, and excise duties on packaging rose), which resulted in a price hike and subsequent decline in sales. Given the recessionary trends in the Estonian economy, there was a general aversion to the introduction of new taxes from retailers and associations (Merilai 2011). Moreover, before the tax reform, several retailers warned of being forced to increase retail prices and pass on the burden of increased costs to suppliers due to tax amendments (Kalmus 2009, Kaljuvee 2010).

Discussion and Conclusions

The dynamics in the Estonian food retail and related industries that were presented in the previous sections are a reflection of the general political economy in Estonia over the last 20 years. The cornerstones of the transition process have been an extreme openness to foreign trade and international capital flows with conservatism in fiscal and monetary policies. Given the high social inequalities, low decommodification level, and non-corporatist attitudes that are revealed in weak labor unions, Estonia has been categorized as a neoliberal welfare state (see Aidukaite 2011, Thorhallsson and Kattel 2012). By liberalizing capital and current account transactions at the beginning of the 1990s, Estonia has served as a low-tax and low-wage investment hub for foreign investors. Given these pull factors, which were accompanied by the global FDI wave in retail as well as the ICT-led transformation of economies in the late 1990s, the Estonian retail industry and related sectors became an attractive destination for Western investors.

Throughout the years, the ensuing developments have been revealed in the drastic transformation of business relations between and within the industries against the contradicting interplay between conservative (nationalistic) consumer preferences and the imported business culture of foreign TNCs (see Chapter 6 of this book). Aside from the established local networks and signs of vertical integration, one could also witness increasing cross-border network embeddedness on the side of manufacturers as well as retailers with a lesser extent of territorial embeddedness which could be explained by the small size of the market. These tendencies are manifest in the reliance on external networks, the increasing share of private label products, and the enclavization processes. The latter have incurred positive feedback mechanisms in the Estonian food industry in terms of prevailing business relations between large companies in different industries. Moreover, the early 2000s, when foreign-owned retail chains established their position in the Estonian food retail market, witnessed an almost 100 percent acquisition of the Estonian banking sector by Nordic financial institutions with the motive of providing financial services to home-country businesses that invested in Estonia. In this way, subsidiaries of TNCs have got preferential access to bank credit in order to finance their investments and rapid expansion in Estonia (see De Haas and Naaborg 2006: 169–74). Therefore, the working of interlinked foreign capital in both financial and non-financial sectors, including food manufacturing and retail, has exposed Estonia to the processes of cumulative causation.

These positive feedback mechanisms in the Estonian food industry have also been revealed in the dualistic structure of the retail sector, that is, niche produce of small enterprises marketed either in open markets or small independent stores with a trivial one to two percent market share, while industrial mass-production dominates in retail chains. Furthermore, the proliferation of a variety of modern retail formats has entailed agglomeration economies which have opened up opportunities for the rejuvenation of small shops in shopping centers or nearby the supermarkets that sell niche farm products. These open markets and medium-sized shops have recently been vigorously developed, with the support by local municipalities. The Estonian Chamber of Commerce and Industry recognizes the popularity of small shops and the trend towards environmentally friendly but more expensive local products, but currently the economic situation favors bigger outlets (Palts 2011).

It is noteworthy that the evolution of the food retail business, led by wholly foreign-owned subsidiaries, has been characterized by the fierce price-based competition among local and foreign retail chains. This, in turn, has laid the ground for aggressive expansion and the active exploration of opportunities to gain higher market share as well as more power and control. In particular, foreign-owned retail chains have put an emphasis on fast growth with increasing investments in fixed assets, which is reflected in relatively high level of debt liabilities compared to local chains (see Figure 5.2).

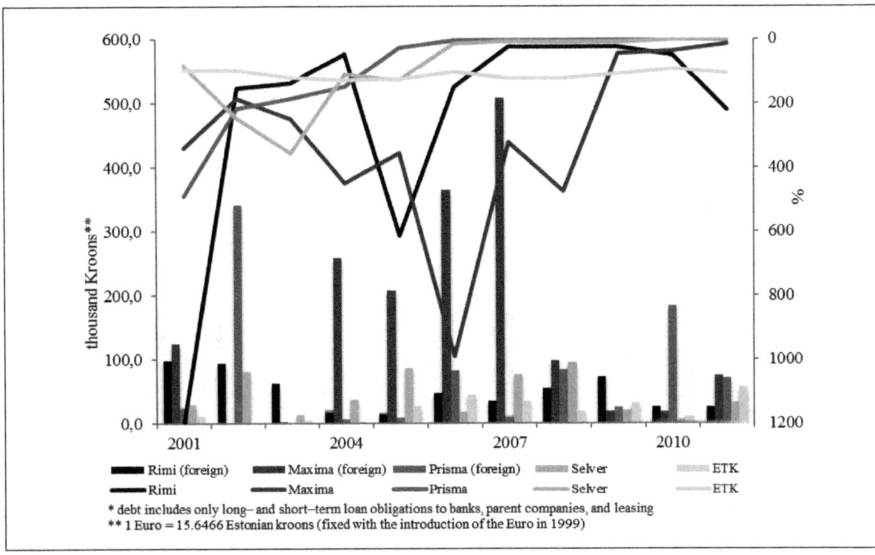

Figure 5.2 Debt-to-equity ratio* (lines; right scale) and gross fixed investments per worker (bars; left scale) of retail chains in Estonia, 2001–2011

Source: Centre of Registers and Information Systems 2012, author's calculations.

On the other hand, contrary to the findings by Baud and Durand (2011), foreign-owned chains have performed less successfully than domestic ones in terms of profitability and solvency indicators (see Figure 5.3), which could explain the tensions arising around the TNCs in the food retail industry.

The business models established in the food retail business have had several implications for general macroeconomic stability and sustainable development. First, multiplier effects in the economy have been undermined by increasing foreign trade, especially cheap imports, and oligopsonistic asymmetrical as well as exploitative business practices that have limited investments in upstream sectors, but also reduced the opportunities to profit from the economies of scale. The vulnerability to the production system stems from the low R&D intensity and outdated productive equipment in the food manufacturing and beverages division, which among other factors, have been affected by the small size of the domestic market as well as retailers' policies (see Pällin 2007, Oppar 2006, Saron 2011). Furthermore, the position of local producers has been influenced by the lack of positive intra-industry spillover effects (Kalvet 2010: 142–143). These tendencies have been reflected in the *primitivization* of the industrial base as well as the de-linking processes that undermine local demand while being further aggravated by the retailers' reliance on global production networks, private labels, or lack of adjustment to Estonian market peculiarities (Sõrmus 2011).

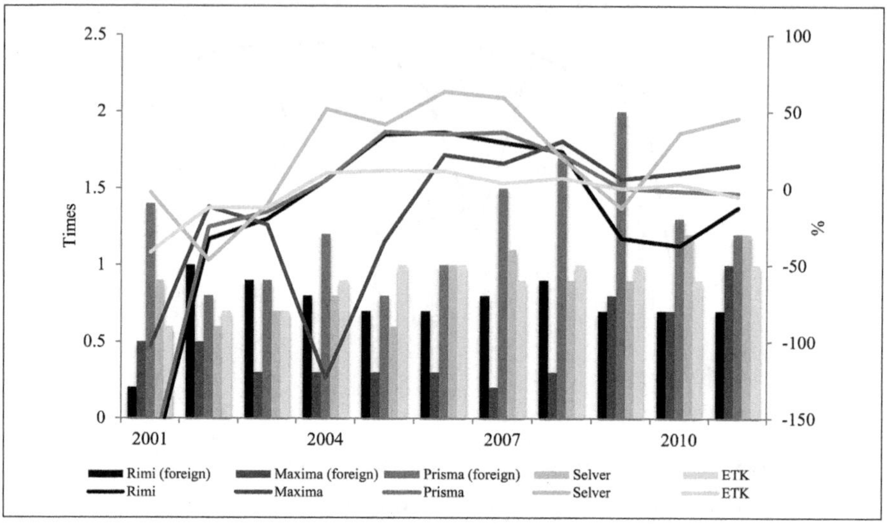

Figure 5.3 Return on equity (lines; right scale) and current ratio (bars; left scale) of retail chains in Estonia, 2001–2011

Source: Centre of Registers and Information Systems 2012, author's calculations.

Second, the reciprocal interdependence between retail chains and large producers has increased systemic risks due to the established business relations. The problems lie in unbalanced market relations and income distribution that have increased the vulnerabilities in the interrelated food production, R&D, distribution, and marketing, where backward and forward linkages have been created over many years: the local food industry cooperates with the producers of added ingredients, packaging, and equipment, while product development is undertaken with R&D establishments. Therefore, the locked-in effect in terms of "too-big-to-fail" exists, that is, increased fragility in the economy due to the concentration and increasing presence of foreign capital in interrelated banking, logistics, retail, and food manufacturing. One could observe the *Kaleckian* principle of increasing risks in terms of a higher reliance on the external financing of investments and growth (see Lavoie 2009). Furthermore, the economy has become vulnerable to pass-through effects, stemming from the dependence on decisions taken and developments occurring abroad. As a transplant region (Jacobs 1984) or a satellite platform (Markusen 1999b) for Scandinavian enterprises, and occupying a fraction of the international value chain of TNCs, Estonia has undermined its resilience with severe difficulties in creating reproductive as well as self-sufficient industries.

We aimed to contribute to the understanding of the interplay between external and internal forces of change that manifests itself when the internationalization of retail and the adoption of a neoliberal political regime meet within the Estonian context. Given the push and pull factors for retail internationalization, Estonia has witnessed the increasing penetration of foreign companies into the food retail industry as well as the gradual market concentration into the hands of six major retail chains, while similar tendencies could be witnessed in farming, manufacturing, and other related businesses. Concentration in all layers of the value chain has been manifest in the movement towards larger enterprises, as predicated in the heterodox economic theory. In a monetary production system, these processes of concentration represent the private sector's attempt to control prices in order to maintain sufficient profits and the ability to undertake further investments (see Minsky 2008). Thus, from a theoretical perspective, the Estonian case displays precisely these outcomes, what could be expected from the introduction of a capitalist economic order with limited government intervention, and, in particular, without the active management of FDI, the central mechanism of accumulation compels firms to grow, gain more control, and increase market share in order to survive. The Estonian retail experience is an example of a short, transitory episode of pure competition followed by the present state of imperfect competition with increased interdependencies, but also imbalances. Thus, when analyzing the impact of FDI on food retail and related economic activities, parallels can be drawn between the Estonian experience over the last 15 years and international decades-long practice, while exceptions can be ascribed to the small size of the market, Estonian political economy, and cultural peculiarities. Local consumers have to some extent managed to constrain the activities of foreign retail chains and push through their preferences for local food with grass-root initiatives in terms of different campaigns and petitions by the civil society groups in order to countervail imports, promote local food produce, and increase the vitality of local producers. The government, though, has mainly focused on promoting higher economic growth, while disregarding the effects of growth on the allocation of resources and distribution of income. Thus, the negative outcomes of the cyclicality and the economy's explosiveness have not been alleviated by government interventions (see Minsky and Ferri 1991). The macroeconomic environment, that is, general fiscal and monetary measures, has instead contributed to the suffocation of the whole economy from the above mentioned "Dutch disease" syndrome and increased instability created by the debt-financed investments during a boom period that unraveled into the 2008 crisis.

Acknowledgments

The research for this article was partially funded by the Estonian Research Council (grant no 8418).

References

Aidukaite, J. 2011. "Balti heaoluriik" pärast kahtkümmet muutuste aastat. *Eesti Inimarengu Aruanne, 2010/2011. Inimarengu Balti rajad: muutuste kaks aastakümmet* [Online]. Eesti Koostöö Kogu, Tallinn, 71–5. Available at: http://kogu.ee/public/eia2011/EIA_2011.pdf [accessed: 24 October 2012].

Aoyama, Y. 2007. Oligopoly and the structural paradox of retail TNCs: an assessment of Carrefour and Wal-Mart in Japan. *Journal of Economic Geography*, 7(4), 471–90.

Baud, C. and Durand, C. 2011. Financialization, globalization and the making of profits by leading retailers. *Socio-Economic Review*, 10(2), 241–66.

Blythman, J. 2005. *Shopped. The Shocking Power of British Supermarkets*. Harper Perennial, London, New York, Toronto and Sydney.

Christopherson, S. 2007. Barriers to "US style" lean retailing: the case of Wal-Mart's failure in Germany. *Journal of Economic Geography*, 7(4), 451–69.

Coe, N.M. and Wrigley, N. 2007. Host economy impacts of transnational retail: the research agenda. *Journal of Economic Geography*, 7(4), 1–32.

Dawson, A.J. 2007. Scoping and conceptualising retailer internationalisation. *Journal of Economic Geography*, 7(4), 373–97.

De Haas, R. and Naaborg, I. 2006. Foreign banks in transition countries: to whom do they lend and how are they financed? *Financial Markets, Institutions & Instruments*, 15(4), 159–99.

Dicken, P. 2011. *Global Shift. Mapping the Changing Contours of the World Economy*. 6th Edition. London: SAGE Publications.

Drahokoupil, J. 2007. From national capitalisms to foreign-led growth, in *Dollarization, Euroization and financial instability: Central and Eastern European countries between stagnation and financial crisis?*, edited by J. Becker and R. Weissenbacher. Marburg: Metropolis Verlag, 87–108.

Drahokoupil, J. 2008. *Globalization and the State in Central and Eastern Europe. The Politics of Foreign Direct Investment*. London: Routledge.

Gallagher, K.P. and Zarsky, L. 2007. *The Enclave Economy. Foreign Investment and Sustainable Development in Mexico's Silicon Valley.* Cambridge, MA: MIT Press.

Hess, M. 2004. "Spatial" relationships? Towards a reconceptualization of embeddedness. *Progress in Human Geography*, 28(2), 165–86.

Humphrey, J. 2007. The supermarket revolution in developing countries: tidal wave or tough competitive struggle. *Journal of Economic Geography*, 7(4), 433–50.

Jacobs, J. 1984. *Cities and the Wealth of Nations. Principles of Economic Life.* New York: Random House.

Kalvet, T. 2010. Business models of the largest enterprises in a small country context: the case of Estonia. *Halduskultuur—Administrative Culture*, 11(1), 128–51.

Keen, S. 2011. *Debunking Economics—Revised and Expanded Edition: The Naked Emperor Dethroned?* 2nd Edition. London, New York: Zed Books.

Lavoie, M. 2009. *Introduction to Post-Keynesian Economics.* 2nd Edition. Basingstoke, UK: Palgrave Macmillan.

Markusen, R.A. 1999a. Four structures for second tier cities, in *Second Tier Cities. Rapid Growth beyond the Metropolis*, edited by Ann R. Markusen, Y.-S. Lee, and S. DiGiovanna. London, Minneapolis: University of Minnesota Press, 21–42.

Markusen, R.A. 1999b. Studying regions by studying firms, in *Second Tier Cities. Rapid Growth beyond the Metropolis*, edited by Ann R. Markusen, Y.-S. Lee, and S. DiGiovanna. London, Minneapolis: University of Minnesota Press, 43–64.

Minsky, P.H. 2008. *Stabilizing an Unstable Economy*. New Haven, CN: Yale University Press.

Minsky, P.H. and Ferri, P. 1991. *Market Processes and Thwarting Systems*. Working Paper No. 64 [Online]. Annandale-on-Hudson: Levy Institute of Bard College. Available at: http://www.levyinstitute.org/pubs/wp64.pdf [accessed: July 2, 2012].

Munson, C.L., Rosenblatt, M.J., and Rosenblatt, Z. 1999. The use and abuse of power in supply chains. *Business Horizon*, 41(1), 55–65.

Oppar, A. 2006. Sector study: manufacture of food products and beverages, in *Private Sector R&D in the New Member States: Case of Estonia*, edited by T. Kalvet. Brussels: European Techno-Economic Policy Support Network.

Reardon, T., Henson, S. and Berdegue, J. 2007. "Proactive fast-tracking" diffusion of supermarkets in developing countries: implications for market institutions and trade. *Journal of Economic Geography*, 7(4), 399–431.

Reardon, T. and Hopkins, R. 2006. The supermarket revolution in developing countries: policies to address emerging tensions among supermarkets, suppliers and traditional retailers. *The European Journal of Development Research*, 18(4), 522–45.

Simms, A. 2007. *Tescopoly. How One Shop Came Out On Top And Why It Matters*. London: Constable.

Thorhallsson, B. and Kattel, R. 2012. Neo-liberal small states and economic crisis: lessons for democratic corporatism. *Journal of Baltic Studies* [Online], 1–21, iFirst. Available at: http://uni.hi.is/baldurt/files/2012/10/Neo-liberal-small-states-Estonia-Iceland-by-Thorhallsson-Kattel.pdf [accessed: November 14, 2012].

Tregear, A. 2011. Progressing knowledge in alternative and local food networks: critical reflections and a research agenda. *Journal of Rural Studies,* 27, 419–30.

Weatherspoon, D.D. and Reardon, T. 2003. The rise of supermarkets in Africa: implications for agro-food systems and the rural poor. *Development Policy Review*, 21(3), 333–55.

Wrigley, N. and Lowe, M. 2002. *Reading Retail*. New York: Arnold; London: Oxford University Press Inc.

Newspapers, Reports and Interviews

Äripäev. 2009. Leedu president vetostas kaubanduse marginaale piirava eelnõu. *Äripäev,* March 17, 2009 [Online]. Available at: http://www.ap3.ee/?PublicationId=3ce335cc-d177-4d2f-8db8-bf654179d50c [accessed: August 2, 2011].

Bank, K. 2010. Tootjad: jaeketid käituvad kui diktaatorid. *Äripäev,* October 16, 2010 [Online]. Available at: http://www.ap3.ee/?PublicationId=c883a903-832f-4d84-a868-74492d3d41ee [accessed: October 10, 2011].

EIER. 2010. *Eesti toidukaupade positsioon siseturul. Kodumaiste ja importtkaupade osakaal Eesti kaupluse käibes ja sortimendis ning hinnad.* Tallinn: Estonian Institute of Economic Research.

EPLO. 2004. VP Market tahab ühe aastaga Eesti kaubanduse esikolmikusse tõusta. *Eesti Päevaleht,* October 7, 2004 [Online]. Available at: http://www.epl.ee/artikkel/275887 [accessed: September 25, 2011].

Erilaid, E. 2004. Odavad Leedu munad vallutavad Eesti turgu. *Postimees,* June 7, 2004 [Online]. Available at: http://www.postimees.ee/170604/esileht/majandus/136327.php [accessed: September 2, 2011].

Eurostat. 2011. EU-Agricultural census 2010—first results. The number of agricultural holdings in the EU27 fell by 20% between 2003 and 2010, while the agricultural area decreased by only 2%. Eurostat News Release, 147/2011. Eurostat Press Office.

Johanson, K. 2007. Väikepoed võitlevad väljasuremisohuga. *Sakala*, December 13, 2007 [Online]. Available at: http://www.tarbija24.ee/151207/esileht/olulised_teemad/tarbija24/kasu/301237.php [accessed: August 22, 2011].

Kaljuvee, A. 2007. Prisma juht: suured tegijad suruvad väikepoed selili. *Eesti Päevaleht*, March 28, 2007 [Online]. Available at: http://www.epl.ee/artikkel/379883 [accessed: May 25, 2011].

Kaljuvee, A. 2010. Maxima lisaks müügimaksu tarnijatele. *Eesti Päevaleht*, July 28, 2010 [Online]. Available at: http://www.epl.ee/artikkel/580867 [accessed: June 18, 2011].

Kallas, R. 2009. Prisma: pakume sama palju Eesti kaupa kui kohalikud ketid. *Eesti Päevaleht*, January 15, 2009 [Online]. Available at: http://www.epl.ee/artikkel/455132 [accessed: April 12, 2011].

Kallas, R. 2011. Margus Venelaine: omatooted labidast lihani? *Delfi Majandus*, March 17, 2011 [Online]. Available at: http://majandus.delfi.ee/news/tarbija/margus-venelaine-omatooted-labidast-lihani.d?id=42268853 [accessed: August 22, 2011].

Kalmus, K. 2009. Ketikaupluste dilemma: kas tõsta hindu ainult Tallinnas või üle Eesti? *Eesti Päevaleht*, December 10, 2009 [Online]. Available at: http://www.epl.ee/artikkel/484572 [accessed: June 28, 2011].

Karotamm, L. 2008. Maxima: eikusagilt kõrgustes. *Postimees*, March 21, 2008 [Online]. Available at: http://www.postimees.ee/150508/esileht/majandus/318902.php [accessed: April 11, 2011].

Kirsipuu, M. 2006. RIMI avas Eestis Baltimaade suurima keskköögi. *Eesti Päevaleht*, September 5, 2006 [Online]. Available at: http://www.epl.ee/artikkel/353528 [accessed: May 10, 2011].

Lindpere, M., Soosaar, O., Pungas, K. and Lambing, M. 2011. Kuidas Eesti toiduaineturg turuosalisi teenib? *Rahapoliitika ja Majandus* [Online], 1. Available at: http://www.eestipank.info/pub/et/dokumendid/publikatsioonid/seeriad/ylevaade/_2011_01/turg.pdf?ok=1 [accessed: September 20, 2011].

Lohk, M. 2008. Prismast kodumaist õuna veel ei saa. *Postimees*, August 29, 2008 [Online]. Available at: http://www.tarbija24.ee/020908/esileht/olulised_teemad/tarbija24/toit/341228.php [accessed: August 7, 2011].

Lohk, M. 2009. Selver lubab pakkuda rohkem soodsaid tooteid. *Postimees*, February 20, 2009 [Online]. Available at: http://www.tarbija24.ee/?id=85030 [accessed: September 30, 2011].

Maasalu, S. 2010. Riik toetab põllumeeste koondumist tootjarühmadesse. Postimees, May 2, 2010 [Online]. Available at: http://www.e24.ee/?id=257431 [accessed: August 12, 2011].

Männiste, A. 2006. Maxima pakub odavamaid hindu. *Postimees*, September 21, 2006 [Online]. Available at: http://www.tarbija24.ee/280906/esileht/olulised_teemad/tarbija24/toit/219124.php [accessed: June 20, 2011].

Merilai, M. 2011. *Interview*. Estonian Traders Association. Transcription: June 14, 2011.

Neudorf, R. 2002. Soome ettevõtete klubi kontrollid Eesti turgu. *Eesti Päevaleht*, June 14, 2002 [Online]. Available at: http://www.epl.ee/artikkel/207643 [accessed: March 10, 2011].

Niitra, S. 2009. Rimi alustas koostööd Eesti mahetootjaga. *Postimees*, August 14, 2009 [Online]. Available at: http://naine.postimees.ee/?id=152690&redir= [accessed: April 25, 2011].

Ninn, E. 2009. Kodumaiste puu- ja köögiviljade osakaal on Rimis tõusnud. *Postimees*, November 2, 2009 [Online]. Available at: http://www.tarbija24.ee/?id=182667 [accessed: June 13, 2011].

Pällin, P. 2007. Toiduainetööstus, in *Kaupmeeste riik*, edited by M. Tiits. Tartu: Eesti Teaduste Akadeemia, 64–77.

Palts, M. 2011. *Interview*. Estonian Chamber of Commerce and Industry. Transcription: May 4, 2011.

PM Online. 2002. Suurpoed ei taha väikeste Eesti firmade kaupa. *Postimees*, July 15, 2002 [Online]. Available at: http://vana.www.postimees.ee/index.html?number=919&op.=lugu&id=64512 [accessed: August 24, 2011].

Potisepp, S. 2006. Sirje Potisepp: Mahlakampaania on eesti toidule vajalik. *Postimees*, October 31, 2006 [Online]. Available at: http://www.postimees.ee/311006/esileht/arvamus/226066.php [accessed: November 29, 2011].

Potisepp, S. 2010. Sirje Potisepp: Eesti liha toetuseks. *Postimees,* March 30, 2010 [Online]. Available at: http://www.e24.ee/?id=243620 [accessed: May 13, 2011].

Potisepp, S. 2011. *Interview*. Association of Estonian Food Industry. Transcription. July 1, 2011.

Raig, T. 2011. Miks jaeketid ühe enam oma logistikat ise korraldavad? *Äripäev*, January 21, 2011 [Online]. Available at: http://www.logistikauudised.ee/article/2011/1/21/miks-jaeketid-uha-enam-oma-logistikat-ise-korraldavad [accessed: April 28, 2011].

Raudoja, T. 2007. Väikese toidupoe vaevad. *Tartu Postimees*, June 12, 2007 [Online]. Available at: http://tartu.postimees.ee/120607/tartu_postimees/arvamus/265722.php [accessed: March 9, 2011].

Reimer, A. 2009. Väikepoed sünnivad taas-esinduslikes kaubakeskustes. *Eesti Päevaleht*, March 18, 2009 [Online]. Available at: http://www.epl.ee/artikkel/462385 [accessed: June 20, 2011].

Reimer, A. 2010. Prisma pakub ennast Eesti toodete eksportijaks Põhjamaadesse. Eesti Päevaleht, May 5, 2010 [Online]. Available at: http://www.epl.ee/artikkel/576191 [accessed: August 22, 2011].

Ronk, A. 2004. Läti ja Leedu odavad kaubad ründavad jõuliselt Eesti poelette. *Eesti Päevaleht*, July 26, 2004 [Online]. Available at: http://www.epl.ee/news/majandus/lati-ja-leedu-odavad-kaubad-rundavad-jouliselt-eesti-poelette.d?id=50989167 [accessed: October 28, 2011].

Saron, T. 2011. *Interview*. Estonian Dairy Association. Transcription: June 14, 2011.

Sokk, E. 2011. *Interview*. Estonian Ministry of Agriculture. Transcription: July 13, 2011.

Sõrm, M. 2011. *Interview*. Estonian Ministry of Economic Affairs and Communications. Transcription: July 18, 2011.

Sõrmus, R. 2011. *Interview*. Estonian Chamber of Agriculture and Commerce. Transcription: May 26, 2011.

Taube, A. 2005. Jaekaubanduses saab kord majja. *Tartu Postimees*, June 13, 2005 [Online]. Available at: http://tartu.postimees.ee/150605/tartu_postimees/168663.php [accessed: April 19, 2011].

Villak, H. 2008. Fazer Amica alustas toodete müüki Prisma Peremarketitesse. *Eesti Päevaleht*, June 25, 2008 [Online]. Available at: http://www.epl.ee/artikkel/433587 [accessed: November 30, 2011].

Chapter 6
Twenty Years of Restructuring in the Estonian Food Retail Industry—the Interplay of the Political Economy of Estonia and Institutional, Historical and Geographical Factors

Egert Juuse, Sylvi Birgit Endresen, Rainer Kattel

Introduction

After the collapse of the Soviet Union in 1991, Estonia faced the challenge of instituting a capitalist market system with a supportive institutional-legal framework. Transformations in the economy were paralleled with the changes in the food retail sector by the gradual introduction of modern retail practices. More importantly, the subsidiaries of retail transnational companies (TNCs) can be considered catalysts of major shifts as from the late 1990s foreign retailers have set the path for the local retail market. By considering the foreign companies in the Estonian food retail market, the current study maps the dynamics in the food retail sector by presenting the key features of the Estonian political economy and the most significant events that have set the trajectory of the development in food retail. Hence, the objective is to present the evolution of the Estonian food retail industry and elaborate on critical junctures, that is, institutional, economic, and political factors that have affected the transformation of the industry. The significance of the study stems from the fact that the introduction of modern retail in Estonia occurred within a very short time period compared to the decades-long transformation of retail in the countries in the West.[1] Moreover, the study offers an opportunity to understand the impact of neoliberal policies, in particular openness to foreign capital and trade, on the retail industry in a small transition economy. Aside from the country's affiliation to the neoliberal policy toolbox which provides a window of opportunity for understanding the outcomes of certain political decisions, FDI in food retail has been an understudied field despite the numerous analyses in the field of FDI in Estonia.[2] Thus, this study offers an opportunity to

1 See Wrigley, Lowe (2002) on the historical development of modern retail.

2 See for instance Tiits (2007), Masso, Roolaht and Varblane (2010), Varblane (2010).

understand the impact of neoliberal policies, in particular the openness to foreign capital and trade, on the retail industry in a small transition economy.

We draw upon the historical institutional perspective (see Pollitt 2012: 39) that enables us to gain insights on the inward internationalization of the Estonian food retail industry and its overall reorganization during the post-Soviet period. In spite of the internationalization aspect of the study which could be addressed with the theoretical approach of *retail internationalization* in terms of furnishing the notion of internationalization in this sub-industry (Alexander and Myers 2000) or analysis of factors affecting the internationalization process (Vida and Fairhurst 1998, Evans et al. 2008), we emphasize the most important indigenous developments that have affected the transformation of the food retail industry, including its internationalization, during the last 20 years.

The chapter is structured as follows: firstly, we give a brief overview of the development of the Estonian food retail industry from the early 1990s, including the introduction of modern food retail and its impact on local independent stores as well as open food markets. Secondly, the impact of FDI on retail transformation and on consumption patterns is elaborated upon by drawing on the understanding and perceptions of key stakeholders within the Estonian business community. Finally, the discussion and concluding remarks convey our understanding of the processes that have shaped the pattern of retail business today, focusing on the interplay of FDI dynamics and weak regulation that explain the concentration of capital and domination of foreign retail chains. With regard to the methods, empirical data has been collected by using statistical information and secondary data, that is, previous surveys and journalistic literature, as well as interviews conducted with key stakeholders in 2010–2011: the Estonian Chamber of Commerce and Industry, the Estonian Traders Association, the Association of Estonian Food Industry, the Estonian Chamber of Agriculture and Commerce, the Estonian Dairy Association, the Estonian Ministry of Economic Affairs and Communications, and the Estonian Ministry of Agriculture.

The Transformation of the Food Retail Industry during the Turbulent 1990s

The 1990s and 2000s are two periods in the Estonian retail sector development with diverging, but also converging trends. In contrast to the Soviet political and economic structures, the establishment of new market-based production systems was carried out under the neoliberal agenda. Belief in the self-adjusting market mechanisms was also reflected in the privatization of retail companies. The decision to sell state owned retail companies for cash to strategic investors was motivated by the need to foster entrepreneurship and decentralize economic power (see Terk 1999). In essence, the aim was to create as many independent retailers as possible by dismantling the huge retail complexes of the Soviet period, such as cooperatives and regional service consortia, and by selling individual stores. The privatization process in Estonia entailed the fragmentation of the market

Table 6.1 Retail chains in Estonia (as of June 2013): Formats* and entry to the market

Retail chain	Formats	Establishment/ownership
ETK co-op.	Mostly supermarkets and convenience stores in rural areas + 7 hypermarkets	Established already in 1902, but operated also during the Soviet period; locally owned company.
Selver Ltd.	Mostly supermarkets and 6 hypermarkets in urban areas.	Established in 1995; locally owned company.
Comarket Ltd.	Only convenience stores in 3 largest urban areas.	Established in 2002; locally owned company.
Rimi Eesti Food Ltd.	Operates mostly hard discounter stores, but also supermarkets and hypermarkets in the largest urban areas.	Entered in 1999; wholly foreign-owned subsidiary.
Prisma Peremarket Ltd.	Only 9 hypermarkets in 3 main urban areas.	Entered in 2000; wholly foreign-owned subsidiary.
Maxima Ltd.	Operates mainly supermarkets and convenience stores + 1 hypermarket in urban areas.	Entered in 2001; wholly foreign-owned subsidiary.

Note: * Hypermarkets, supermarkets, convenience stores, and hard discounters differ in terms of the spanning area and the assortment (see Wrigley, Lowe 2002).
Source: own elaboration based on data from retail chains' websites.

and the proliferation of small retailers. Similar dynamics could be observed in wholesale: the number of wholesale companies increased from two before independence in 1991 to 4,980 in 1994 (Voog 2011). While over 90 percent of the companies in services and commerce were state-owned in 1991, 94 percent of retailers and 90 percent of wholesalers were privatized by 1994 (OECD 2000). Such drastic changes took place against the slack regulations and low level of institutional development that explain the emergence of open markets as focal points in food retail at that time. For instance, several rounds of the banking crises in the 1990s and the low level of supportive technological infrastructure in terms of non-existent non-cash payments in the early 1990s prevented the food retail industry from taking off (Voog 2011). However, more important for the long-term development of the food retail business was the concentration of control over the privatized companies into the hands of the top management of enterprises, as the Estonian privatization model did not favor mass privatization such as the takeover of privatized companies by employees. The reason for this was that a significant share of non-native language speakers in the population incurred tensions on the grounds of nationality after regaining independence.

Furthermore, local residents did not have sufficient financial resources to partake in the privatization process, which explains why many small and medium-sized companies in commerce that could serve as a gateway to the Baltic markets were taken over by foreigners. The attractiveness of the Estonian commercial sector for foreign investors became evident; by 1995, half of foreign-owned companies were involved in commerce and trading activities (OECD 2000). In general, the political economy of the Estonian food retail industry was from the outset geared towards establishing many independent retailers in a free and perfectly competitive non-interventionist market economy that was subject to foreign competition and a low level of unionization (see Thorhallsson and Kattel 2012 on the Estonian political economy). Affected by the stabilization of the economy and the adjustment of the Estonian legal-institutional framework in line with international standards, the gradual takeover of manufacturing companies as well as banks by foreign companies prepared the ground for foreign investments in the Estonian food retail (and other service sectors).

The first foreign capital based retail chain *REMA1000* launched its business operations in Estonia in 1995; and a year later, the first stores under the *SPAR* brand name were opened. In spite of their expansion plans in both the retail and wholesale segments (see Pirn 1997, Tomak 1997a), there was an even division of the market between open markets and grocery stores (Murde 1995). Moreover, only about *one quarter* of the grocery and convenience stores were operating in chains in the mid-1990s. These chains were local ones—*ETK* and *A-Selver*—and foreign owned—*SPAR* and *REMA1000* (Tomak 1997b). The rest of the market was shared between smaller convenience stores and food stalls in open markets. Thus, the second part of the 1990s was still favorable for local businesses in retail and production due to relatively low competition from foreigners. Yet, the consolidation among retailers as well as organic expansion of retail chains intensified cost-based competition and triggered the gradual demise of independent small shops that was spurred on by ill-advised business strategies and management failures. The economic reason behind the demise of independent retailers and the expansion of chains was found in the workings of economies of scale that enabled the chains to outcompete small shops with lower prices. Moreover, retail chains also had a better bargaining position against producers and vendors as compared to independent stores and could therefore keep costs lower (Taube 2005). Other factors behind the closure of small shops were stringent regulations in terms of hygiene, safety, and other norms that required compliance with European Union (EU) directives (Hint 2002); but also a virtual lack of local land-use planning. Of the idiosyncratic contributors to the worsening situation of retailers were the 1997–1998 economic crises that forced some of the retailers and wholesalers to leave the market or merge with each other. Given these developments, on average 4.5 independent shops were closed per day and in total 1,600 units disappeared during a two-year period from 2000 to 2001 (Neudorf 2002). In general, tightening competitive and regulatory pressures displaced small shops and open markets with supermarkets, hypermarkets, hard discounters as well as shopping centers that offered lower prices, a wider assortment of products, and longer opening hours. Independent retailers were conceived of as mainly the intermediaries of fresh farm products (Tigasson 2002).

Yet, foreign owned retail chains were struggling in the 1990s as well, which led to the collapse of *SPAR Estonia*—the only exit of a foreign retailer from the Estonian market. By 1999 it was evident that bad investment decisions and indebtedness indicated financial problems and forced the owners of *SPAR Estonia* to sell the chain to its main foreign competitor (Erilaid 1999). Nonetheless, *SPAR Estonia* collapsed in 2001 because of the conflicts and misunderstandings between the owners and management, highly leveraged investments, and increasing indebtedness to suppliers (EPL 2001). After the bankruptcy in 2002, liabilities were obtained by the local company *ABC Group*, which turned former *SPAR* stores under a new concept into a retail chain *Comarket* that has been operating ever since (ETA 2002). The case of *SPAR Estonia* shows clearly how financial problems in one of the largest retail chains were transmitted to banking and other sectors, reflecting a *Minsky* moment,[3] where highly leveraged positions with inadequate cash flows increased systemic risks.

The Expansionist 2000s

Notwithstanding the problems of *SPAR Estonia*, the 2000s mark the advent of aggressive market expansion of Nordic and Lithuanian retail chains in Estonia. Between 2002 and 2010 the FDI stock in Estonian retail, wholesale, and vehicle maintenance increased by around 300 percent from 477.5 to 1403.6 million Euros (Bank of Estonia 2011). By 2010, three foreign and three local capital based chains dictated the market with approximately 80 percent of the market share (see Figure 6.1).

Geographic Expansion and Diversification of Format

With regard to geographic expansion, retail chains followed a hierarchical *diffusionist* pattern, starting from the capital city Tallinn, moving to regional centers, and eventually occupying small town markets (Baltic News Service 2001, Riikoja 2006). The inflow of foreign capital coincided with the urbanization process that contributed to the demise of small shops in rural areas, implying the direct link between the retail transformation and urbanization (Palts 2011). The Finnish *Prisma Peremarket Ltd.* operated only in Tallinn for 10 years with five hypermarkets, before moving to Tartu and Narva in the eastern part of Estonia. The exception has been the Lithuanian *Maxima Ltd.* which entered the Estonian retail market in 2001 through the city of Pärnu in southern Estonia with a hard discounter format, but later engaged in super- and hypermarket formats. A peculiarity of the Lithuanian chain has been

3 See Minsky (2008) for his financial theory of investment as well as investment theory of business cycles and their relatedness to financial fragility.

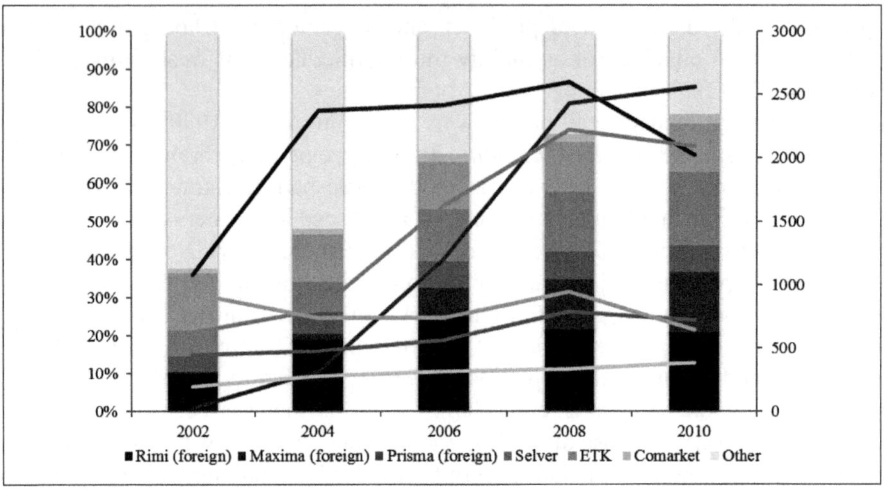

Figure 6.1 The share of net sales (bars; left scale) and the number of employees in the six largest retail chains (lines; right scale) in Estonia, 2002–2010

Source: own elaboration based on data from Centre of Registers and Information Systems (2012).

its policy to target geographical areas where competition is slack such as the eastern parts of Tallinn and Estonia. In these areas, dominated by the Russian speaking population, consumers have been very price sensitive and receptive to imported goods (see Karotamm 2008). As seen in Figures 6.2 and 6.3, the introduction of modern retail with diverse formats has significantly affected the landscape of urban areas. The two largest cities, Tallinn and Tartu, where approximately 40 percent of the Estonian population resides, have witnessed the explosive development of 51 new hypermarkets and supermarkets by local and foreign retail chains after 2000, amounting to a total number of 63 of these formats in 2012.

Increasing competitive pressures have been the incentive for such a rapid expansion. During the first period there was an emphasis on hypermarket formats and shopping centers that primarily expanded through greenfield investments. The next round of expansion consisted mostly of foreign as well as local capital based supermarket and discounter chains (see Roonemaa 2003). The expansion of supermarket and discounter chains has been carried out through mergers and acquisitions as well as organically. Furthermore, retail chains have vigorously diversified into deli, pastry, and bakery businesses as well as engaging in insurance and travel brokerage, and are considering expansion into pharmacy and postal services (Erala 2009b, Merilai 2011). More importantly, the financing of expansion has varied among retailers. While *Prisma Peremarket Ltd.* has relied on accumulated retained earnings, *Maxima Ltd.* has used external sources

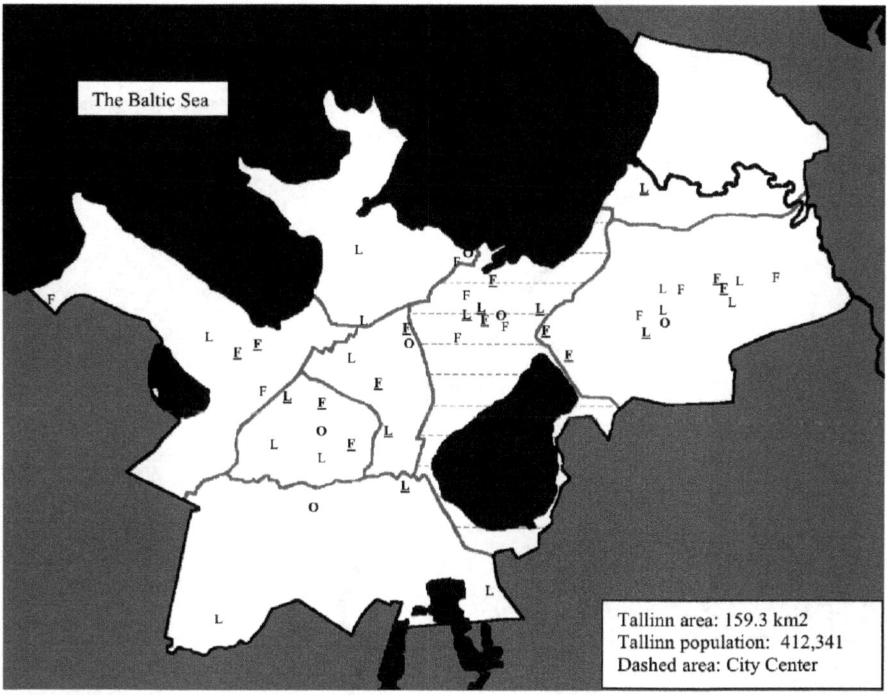

Figure 6.2 Location of open markets, super- and hypermarkets, and shopping centers in the capital city, Tallinn, 2012[*]

Note: Key: L – Local Hypermarket/Shopping Center; L – Local Supermarket; F –Foreign Hypermarket/Shopping Center; F – Foreign Supermarket; O – Open Markets. [*] Does not include 18 *Maxima* convenience stores, 16 *Rimi* convenience stores and 9 *Comarket* convenience stores.
Source: own elaboration based on data from Tallinn official website (2012), Mmh (2008).

(Koovit 2009). As covered in Chapter 5, compared to competitors, *Prisma Peremarket Ltd.*, among the foreign chains, and local chain *Selver Ltd.* have been the most financially sound food retailers with sufficient margins of safety, which indicates the sustained ability to service liabilities and rely on retained earnings in financing investments. All in all, it was during the 2000s that retail chains gained leverage in the value chain of producer–wholesaler–retailer–consumer, whereas shopping centers in all major cities in Estonia were turned into the centers of commerce.

Nationality of Owners and the Impact of FDI

Given the importance of cultural factors for the successful expansion of retail chains that has been studied by Aoyama (2007) and Christopherson (2007), the expansion of Lithuanian, Swedish and Finnish retail chains in Estonia can be explained by the cultural proximity of investor and host countries, which makes

Figure 6.3 Location of super- and hypermarkets, and shopping centers in Tartu, 2012*

Note: Key: L – Local Hypermarket/Shopping Center; L – Local Supermarket; F –Foreign Hypermarket/Shopping Center; F – Foreign Supermarket; O – Open Markets. * Does not include 5 *Maxima* convenience stores, 4 *Rimi* convenience stores and 4 *Comarket* convenience stores.
Source: own elaboration based on data from Alts (2012), Tartu Linnavalitsus (2012).

it possible to overcome the trade-off between standardization and localization. There are, according to the Estonian Traders Association, peculiarities in Estonia, since local chains thrive too, unlike in other CEE countries (Merilai 2011).

As discussed in Chapter 5, the blending of different business cultures in a competitive environment has to some extent standardized the service provision and business relations in the Estonian retail industry. Estonian *Selver Ltd.* and Scandinavian retail chains have managed to set best practice standards in the retail business which has caused troubles for Lithuanian *Maxima Ltd.* For years, the media has published cases of law infringements by *Maxima Ltd.*, mainly related to insufficient command of the Estonian language by their personnel, a lack of or misleading information about products, violations concerning the price tags, and troublesome employment relations (Lohk 2007, Mallene 2011). Supported by the media, Estonian civil society concerned about the undesirable general effects of the chains' expansion has placed most of the complaints at *Maxima's* doorstep (Seaver 2009). By capitalizing on the frayed reputation of some of the foreign retailers, the local retail chain *Selver Ltd.* has managed to create a positive image among consumers. The company underlines that they follow the principles of corporate social responsibility by having programs for its workers and supporting sports activities (Niitra 2010). Given the problems surrounding the Lithuanian retail chain, the lack of interest for investments in Estonia from non-Nordic countries, including the Russian Federation, could be explained by the cultural peculiarities, very small and saturated local market, and legal-institutional differences.

One of the consequences, related to foreign ownership, has been an increased propensity of foreign owned retail chains to prefer to have imported goods instead of local products on their shelves, which is reflected in Figure 6.4 below.

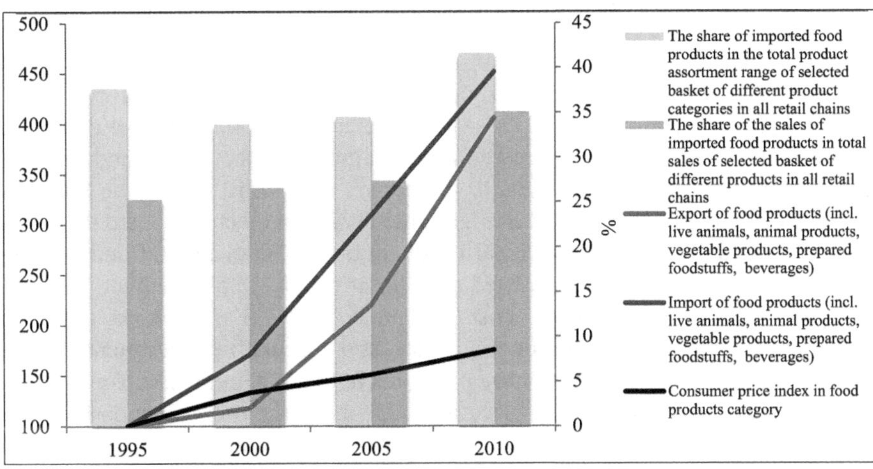

Figure 6.4 The share of imported food products in retail chains (bars; right scale) and foreign trade of food products and consumer price index (lines; index 1995=100; left scale) in Estonia, 1995–2010

Source: author's calculations based on data from Statistics Estonia (2011), EIER (2003, 2010).

In spite of the detrimental effect of the increasing reliance on imports on local producers, a major stakeholder, the Estonian Traders Association, does not see any problems with the operation of foreign retail chains in Estonia and even welcomes the inflow of FDI in retail. Competition leads to the upgrading of local shops, increases efficiency and quality, and external financial support provides buffers during difficult times (Merilai, 2011).

Private Label Products as Overt and Covert Imports

An innovative tool for gaining larger market share has been the introduction of private label products, that is, products marketed under the retailers' own brands. The number of these products has significantly increased in Estonia, as this is a direct consequence of FDI (Sõrmus 2011). In 2002, one percent of all products were private labels, while in 2010 the same figure was eight and reaching 30 percent in the category of dairy products (EMOR 2010). Over the years, private labels have extended into different product categories, including both food and non-food products (Erala 2009a). Their increased popularity could be explained by the increased price sensitivity of local consumers, as lower prices are achieved due to the insignificant marketing costs and scale economies (Kallas 2011). More importantly, private label products tend to be imported (EIER 2010). For instance, most of the *Prisma Peremarket Ltd.'s* private label items are imported due to bottlenecks in the local supply (Villak 2010). It is argued that Estonian food manufacturers are too small to supply retailers with large quantities of these products (Sõrmus 2011, Palts 2011). The argument is supported by the dynamics seen in Figure 6.4 above: imported goods have increased their share and sales, as the overall prices of food products have been on an upward trend. The Estonian Institute of Economic Research has highlighted two main factors behind the rising prices. First, the prevalence of local products on the retail market and second, the business model of foreign-owned subsidiaries in the retail segment has been based on values other than low costs (Männiste 2006). Furthermore, trends in global market prices, rising production costs, and speculative investments before the crisis of 2008 caused inflationary pressures in the retail market (Josing 2007). An additional factor behind the price hikes has been the role of support service providers. Due to the popularity of credit and debit cards among consumers as a means of payment and the dominance of a few commercial banks, banks have been able to impose a fee as a percentage of turnover, ranging between 1 and 3 percent, related to card payments. This can be seen as an extra value added tax on retailers and thus added to the price of items sold (Reinap 2003). Therefore, the promotion of private label products that counter the price increases has been welcomed by the most price sensitive consumers, especially, given the deteriorating purchasing power that suppressed local demand after the 2008 financial crisis.

On the other hand, the very characteristics of private label products that bring prices down—they tend to be cheaper imports and narrow the assortment—run counter to consumer preferences for food products (see Chapter 5 on the impact of private labels on related industries). It has been found that Estonian consumers tend to have a preference for local products, which are considered to be safer and more reliable (Potisepp 2010). A journalistic survey (Sikk 2009) has found that Estonian consumers value a wide range of low-price and high-quality products that are locally produced. They expect spacious shopping areas, more immediate customer service, freshness, and would like to see more locally-owned retailers. Hence, consumer preferences clearly indicate a positive relationship between the higher share of local food products in the assortment and the improved standard of living, but as Figure 6.4 (above) indicates, inflation and deteriorating purchasing power have increased the popularity of cheaper imported goods. While 70 percent of consumers preferred local products in 2010, worsening economic conditions forced people to shift their purchases towards imports (EIER 2010, Merilai 2011, Potisepp 2011).

Countering the Dominance of Retail Chains

Given the increasing dominance of food retail chains and unfavorable legislative amendments, the outlook for small independent shops has not been positive. Nonetheless, novel retail formats have brought about agglomeration economies and thereby opened up possibilities for the rejuvenation of small shops. These stores have been established either permanently or seasonally in shopping centers or nearby supermarkets and have been selling niche farm products. In some cases, retail chains and shopping centers have organized fairs of their own, where small local producers and farmers may rent an area to sell their products (Tammela 2009, Merilai 2011). This has been made possible due to high charges and requested discounts, complicated logistics, and penalizing fines by retail chains. Hence, getting niche products "listed" by food retail chains is a major barrier. This has motivated small producers to combine forces with small shops to sell niche and exclusive products in a highly competitive market, a recognition shared by the Estonian Ministry of Agriculture (Reimer 2009, Sokk 2011).

Many Estonian consumers would also like to see reverse trends in retail towards more direct contact with sales persons (Saar cited in Salu 2008). This coincides with another trend that has recently emerged, namely locally produced and environmentally-friendly food that is increasingly promoted in the media. The Estonian Chamber of Commerce and Industry believes that this trend will increase in the future (Palts 2011). In addition, local authorities have recently given support to the development of open markets in urban areas. However, they have less than one percent of the market for green products, and only 20 green producers have started to supply directly to consumers (Sõrmus 2011). Furthermore, the peculiarity of open markets in the capital city has been their "politicization" in terms of favoring companies that sponsor political parties in power in the municipality (Gnadenteich 2009). Such a policy has caused disdain and made open markets unattractive. Yet, the large-scale expansion of modern retail in populous urban areas could be considered as the main reason behind the low popularity of open markets. In addition, there are two problematic issues in connection with open markets and small producers who sell goods without a license near to shopping centers; these are food safety and tax evasion (Potisepp 2011). Thus, there tends to be a trade-off between vigorous support for small producers that infringes the equality principle, and lack of responsibility towards consumers and the state.

There has been resistance to the practices of retail chains on the demand side as well. Estonian civil society has been vocal in determining the developments of the retail market by supporting Estonian farmers and demanding local products. For instance, in spring 2010, *Rimi Eesti Food Ltd.* decided to discard Estonian pork and beef with the argument of insufficient supplies as well as relatively high prices, and sell only imported meat under private labels (see Östholm cited in Hankewitz 2010). That decision caused a heated reaction from civil society as well as producers, who called for the boycott of *Rimi* supermarkets, and the case caused an outbreak of public anger, even in social media. The Ministry of Agriculture and local producers considered the low price as the main reason behind *Rimi's* decision to favor imported meat and the step was interpreted as a tactic to pressure local producers (see Seeder cited in Kass 2010, Lohk 2010a). After one week, *Rimi* admitted the mistake and reintroduced Estonian pork and beef. In the aftermath of that scandal, press releases about support of local small producers could be interpreted as the pursuit of rehabilitation—around 30 small producers, mainly farms in the dairy industry, supplied *Rimi* hyper- and supermarkets with diverse niche products (Sõrmus 2011, Erala 2010). Yet, in direct competition to local *green* producers, *Rimi* introduced its own private label in 2010 in the green products category, comprised of organic food produced in Sweden and other foreign countries (Lohk 2010b). Thus, in order to countervail imports, promote local food produce, and increase the vitality of local producers, there have been several campaigns[4] launched during the last couple of years against the increasing dominance of retail chains.

Aside from the grass-root level initiatives, there have not been any legally binding measures to intervene in the food retail business to put the brakes on rapid expansion or concentration. The commerce law from 2004 regulates issues of importance for retail, but currently there is neither a deliberate retail policy nor political discussions aiming to change the status quo. Among other factors, this is due to the restrictions imposed by the international agreements:

> Policy measures that distort the functioning of the free market are not feasible in the EU. In general, the European Union dictates what government can and cannot do. After all, Estonian retail sector is well regulated in terms of competition, contractual arrangements, and labor relations. (Interview, M. Sõrm, Ministry of Economic Affairs and Communication 2011)

Thus, stricter state regulations have not been on the agenda. The authorities see foreign companies in a good light as long as those provide jobs, reinvest profits,

4 Campaigns like *Estonian Flag*, *Eco-Sign*, *Best Food Product*, *Homegrown*, and other.

and pay taxes (Sokk 2011, Sõrm 2011). In general, there is a strong belief in the free market economy among the stakeholders (Merilai 2011, Potisepp 2011). The overall stance is that the economy should be free of government intervention, unless competition law has been infringed (Sõrm 2011).

The only political measures that have affected retailers as well as producers have been related to alcohol policy. These are restrictions on advertisements as well as sale areas and opening hours for the sale of alcohol (Paves 2010). Even though such measures could have confined retailers, regulators succumbed to the threats of retail chains to start closing stores earlier in the evening (Masing 2007). Authorities at the local government level have only occasionally directed or hindered the expansion of retail chains with an argument of contradicting city planning principles as well as health and safety requirements (Merilai 2011, EPLO 2003, Peensoo 2003). As specific and targeted policy measures have been off the agenda, only fiscal policies have been of noteworthy relevance for retailers: 'Economic cycles influence the consumption patterns and affect retailers' turnover the most. Other factors that affect retailers are taxes and demographic situation.' (Interview, M. Merilai, Estonian Traders Association 2011)

Discussion and Concluding Remarks

From 1991 onwards, the food retail industry has undergone profound changes. The first stage of restructuring was marked by the proliferation of independent shops and open markets, while the second stage was characterized by the consolidating tendencies among local and foreign owned retail chains. There are several reasons for such a path-changing evolution, and in the following section we discuss the processes we consider most important when seeking explanations for the patterns observed. Given the small market in Estonia with a population of around 1.3 million, fierce price- and cost-based competition among local and foreign food retail companies laid the ground for active exploration of possibilities for expanding market share as well as gaining more control. The result has been the concentration of the retail market into the hands of six leading retail chains; of these, three are non-local and control about 40 percent of the market. The most general process that we wish to highlight here is a consolidating trend that can be predicated in the heterodox economic tradition whereby market forces induce imperfect competition among firms—oligopolistic markets with interdependent relations. Firms endeavor to increase their size and market share by utilizing the mechanisms of economies of scale. By exploiting the cost advantages, large firms drive small firms out of business and thus render perfectly competitive markets into a situation of either monopoly or oligopoly. Therefore, economies of scale are an important reason why most industries are dominated by a small number of large companies. This, in turn, has created a situation, where the strong either crush or absorb the weak (see Minsky 2008, Lavoie 2009, Keen 2011). From this theoretical vantage point, Estonian experience reflects the central mechanism of accumulation that compels firms to grow in order to survive and gain more power, while a short, transitory state of pure competition in the early 1990s was succeeded by the present state of imperfect competition in oligopolistic markets.

Second, macroeconomic policies lie behind the process of change and explain the described concentration in the food retail market. From the very beginning in the early 1990s, Estonian governments have adhered to the free market (laissez-faire) approach by implementing the so-called *Washington Consensus* policies. For many years, Estonia has endeavored to serve as a low-tax and low-wage investment hub for foreign investors. In that respect, the tax reform of 2000 that exempted undistributed profits from corporate income tax, boosted foreign investments. In addition, relatively expansionary fiscal policies during the 2000s before the 2008 crisis stimulated the growth of domestic consumption, as during the boom years, wage increases in the public sector were coupled with the easing of the general tax burden, all of which contributed to the overheating of the economy (see Bernhardtson, Billborn 2010: 10).

The third process that explains the aggressive expansion of retail chains in urban areas, particularly in metropolitan centers, was the increasing *financialization* of the economy since the late 1990s. The turning point in the development of the Estonian banking system came in 1997–1998, when local commercial banks were taken over by Scandinavian financial institutions. The credit supply was thereby increased due to the access to long-term foreign funding. Essentially, the ratio of the banking system's assets (mostly loans and leasing) to GDP doubled in eight years and reached 132 percent at the beginning of 2008. Thus, 2000s saw an unprecedented credit boom that was made possible due the declining cost of credit on the global scale, Estonia's catching-up process from the initial low-level of bank intermediation, high GDP growth, consumer confidence from rising incomes, and expansionary fiscal policies (Voog 2011). Followed by the real estate sector and private persons, most of the bank loans, allocated by the two largest commercial banks in the early 2000s, were absorbed by retail and wholesale industries (SEB 2011, Swedbank 2011). Hence, given the high growth rate of the overall economy in 2000s that was fueled by debt-led domestic consumption as well as investments, both local and foreign owned food retail chains undertook aggressive expansion that enabled them to attain continuing annual increases in sales throughout these years.

The cumulative impact of the commercial banking activities and the neoliberal environment on the development of the food retail industry has been effectuated through the fourth process that we wish to highlight: the urbanization process from the early 1990s and, in particular, the suburbanization trends in the 2000s (see Tammaru et al. 2009). These trends have been affected by the transformation of the structure of the Estonian economy towards services that spurred migration out of rural areas. Suburbanization has been made possible by the increasing mortgage lending that was non-existent in the 1990s due to difficulties in applying for collateralized bank loans and high interest rates, while low purchasing power contributed to an inactive real estate market (OECD 2000). However, the underlying causes for such deepening interdependencies between commercial banking, real estate development, (sub)urbanization, and the expansion of retail chains were the privatization process and land restitution. This placed property in private hands, and eventually paved the way for urban sprawl with increasing in-fills and multifamily housing constructions around the urban centers (Tammaru et al. 2009). These developments are revealed in the locations of retail stores in Tallinn and Tartu (see Figures 6.2 and 6.3, above), as new hyper- and supermarkets in the two largest cities (but also in other urban areas in Estonia during the late 1990s and 2000s) were established either in the city centers or the surrounding districts in the residential areas within the borders of the cities. Essentially, the investment decisions of retail chains followed the geographical expansion of demand and real estate development that was increasingly financed through credit during the 2000s. To some extent, these findings are in contrast to the general pattern of the spatial location of retail outlets in urban areas of western economies—in the outskirts of cities—in particular, in Anglo-American countries (see Wrigley, Lowe 2002).

Our starting point was the prospect of higher returns on investments that gave an impetus and thus rendered the Estonian economy an alluring destination for Western investors in late 1990s and early 2000s. Cultural and geographic proximity of Nordic (and Lithuanian) retailers cannot be overlooked as a factor behind the expansion into the Estonian market. Aside from strengthening their position, foreign retail chains have managed to set the standards by transferring their operational techniques and business models, which have enabled them to increase the market share and thus influence production systems (see Chapter 5, also Humphrey 2007, and Christopherson 2007 on the international experience). However, as several cases indicate, foreign food retail chains have also struggled in Estonia and have been constantly seeking to adjust to the Estonian socioeconomic environment. Moreover, further dominance of foreign retail chains in the future can be seriously compromised, given the nationalistic mood and the preference for local produce among consumers, the resilience of local chains in withstanding the competitive pressures from foreign TNCs, and the rejuvenation of alternative food stores and open markets. In addition, according to European standards, the Estonian retail market was already saturated in 2002. Rapid commercial property development in the Tallinn area that exceeded the growth of real income by three times between 2000 and 2002 resulted in a supply glut (Kivistik 2002).

Irrespective of these peculiarities in the Estonian market, clear parallels can be drawn between Estonia's 15 year long experience in the food retail industry and international practice, including Central and Eastern European countries, given the common heritage of the planned economies (Knezevic and Szarucki 2001, Michalak 2001). General patterns in the food retail industry on the global scale offer evidence on a rapid expansion of subsidiaries of retail TNCs in terms of: 1) geographic diffusion, retail format, and customer segments, 2) consolidation patterns and the increasing market share of these subsidiaries, 3) introduction of private label products. These effects can be identified in Estonia as well. The oligopolistic position of foreign retail chains has been achieved through the increasing reliance on international supply networks to source the products. This is witnessed by the rapid spread of private label products in Estonia that attests the positive relationship between retail concentration and the intensity of private label products as well as private label supply networks in retail internationalization processes (see Wrigley, Lowe 2002: 64). These developments are supported by the increasing share of imported products that enables retailers to capitalize on price arbitrage in search of operational and network efficiencies (see Dicken 2011: 436). There are several implications of such developments. On the one hand, risks stem from the retailers' stocking policy in terms of sourcing the national brand leaders plus private labels, which lead to the demise of small manufacturers and a consequent reduction of choice (see PM Online 2002, Voog 2011). On the other hand, the potential outcome of increasing market share in the hands of foreign retail companies could be higher prices for consumers due to cost or exchange rate based pass-through effects (see Bloch and Olive 1999). Furthermore, the vulnerabilities in relation to pass-through effects stem from the dependence on decisions and developments that are taken and determined externally. As stated above, the option of government intervention to mitigate these risks has been taken off the table. The same macroeconomic policy environment in terms of conservatism in fiscal policies has been sustained for almost 20 years with the belief in a simple taxation system and balanced government budget for the purpose of price and macroeconomic stability. In practice, such an environment has had a significant effect on the unsustainable credit growth and deepening imbalances in the economy created by debt-financed investments during the boom period that unraveled into crisis and threatened the economy with an eventual debt-deflation process. Thus, the period of tranquility and financial stability before 2008 fed buoyant moods in terms of high profit expectations, which were manifest in higher investment activity in the retail and real estate markets in 2000s (see Chapter 5). With hindsight, the pre-2008 boom period bred instability due to the propensity to hold less in liquid assets and issue more liabilities. Furthermore, the increasing share of FDI in the Estonian food retail industry has had a negative impact on the external balance of the industry in terms of its increasing debt liabilities to foreigners, and contributed to the deteriorating balance of the payment position of Estonia due to continuous FDI inflows, not to mention the negative food trade balance. On top of it, given the shrinking population and volatile purchasing power of consumers, the food retail industry faces excess supply with a saturated local market. In these circumstances, keener competition and mounting accumulation of foreign capital have increased the concentration level and rendered Estonian retail business more fragile. Yet, according to stakeholders, for instance the Estonian Chamber of Commerce and Industry, the strategy of attracting foreign capital has been successful: "FDI has been beneficial for Estonia, while not incurring so much negative aspects. The main impact has been in terms of employment creation. Reliance on FDI is quite inevitable for small countries" (Interview, M. Palts 2011).

We aimed to understand the restructuring of the Estonian food retail industry by drawing upon the international experience of the expansion of retail FDI. For comprehending the dynamics in retail business, key features of the Estonian political economy and crucial events in determining the trajectory of the development in food retail business have been highlighted. As demonstrated, the development path of Estonian food retail has been affected by institutional and economic adjustments such as radical change in the political regime and ideology in the early 1990s, transformation of the structure of the economy and the related urbanization process, acquisition of the banking sector by Scandinavian banks, and the conclusion of the privatization process, while peculiarities could be attributed to the historical legacy of the Soviet period, small market size, and *non-corporate—conservative(neoliberal)—nationalistic* socioeconomic environment. As shown, cultural inertia in society has constrained the retailers' activities in shaping the tastes and implementing business practices. On several occasions, civil society has managed to influence foreign retail chains by pushing through consumers' preferences for local food with different initiatives, such as campaigns and petitions, in order to countervail imports, support local food producers, and hinder the establishment of stores. Yet, in many respects, the Estonian case mirrors general trends in modern retail in the international arena. Thus, from a theoretical perspective, the Estonian case displays the outcomes that could be expected from the introduction of a free market order with limited government intervention and, in particular, without the active management of FDI. Unfettered private interests led spatial planning and development of residential and commercial real estate, determined by the activities of commercial banks, real estate developers, and the preferences of residents and companies. Without significant policy constraints, the central mechanism of private capital accumulation has compelled retail companies to grow, gain more control, and increase their market share.

Acknowledgments

The research for this article was partially funded by the Estonian Research Council (grant no 8418).

References

Alexander, N. and Myers, H. 2000. The retail internationalisation process, *International Marketing Review*, 17(4/5): 334–53.

Aoyama, Y. 2007. Oligopoly and the structural paradox of retail TNCs: an assessment of Carrefour and Wal-Mart in Japan. *Journal of Economic Geography*, 7(4): 471–90.

Bernhardtson, E. and Billborn, J. 2010. The role of the banking system in financial crisis—a comparison between the crisis in Asia and the crisis in the Baltic countries. *Sveriges Riksbank Economic Review*, 3: 5–22.

Bloch, H and Olive, M. 1999. *Cyclical Movements in Markups: The Role of Cost Pass-Through.* School of Economics and Finance, Curtin University of Technology.

Christopherson, S. 2007. Barriers to 'US style' lean retailing: the case of Wal-Mart's failure in Germany. *Journal of Economic Geography*, 7(4): 451–69.

Dicken, P. 2011. *Global Shift: Mapping the Changing Contours of the World Economy*. 6th Edition. London: SAGE Publications.

Evans, J., Bridson, K., Byron, J. and Medway, D. 2008. Revisiting retail internationalization—drivers, impediments and business strategy. *International Journal of Distribution and Management*, 36(4): 260–80.

Humphrey, J. 2007. The supermarket revolution in developing countries: tidal wave or tough competitive struggle. *Journal of Economic Geography*, 7(4): 433–50.

Keen, S. 2011. *Debunking Economics—Revised and expanded edition: The Naked Emperor Dethroned?* 2nd Edition. London, New York: Zed Books.

Knezevic, B. and Szarucki, M. 2001. *Retail Business Development: Internationalization in Croatia, Lithuania and Poland.* 7th International Scientific Conference 'Business and Management 2012,' May 10–11, 2012, Vilnius, Lithuania.

Lavoie, M. 2009. *Introduction to Post-Keynesian Economics.* 2nd Edition. Basingstoke, UK: Palgrave Macmillan.

Masso, J., Roolaht, T. and Varblane, U. 2010. Foreign direct investment and innovation in Central and Eastern Europe: evidence from Estonia. *University of Tartu, Faculty of Economics and Business Administration, Working Paper Series*, 67, 1–67.

Michalak, W.Z. 2001. Retail in Poland: an assessment of changing market and foreign investment conditions. *Canadian Journal of Regional Science*, 24(3): 485–504.

Minsky, P.H. 2008. *Stabilizing an Unstable Economy.* New Haven, CT: Yale University Press.

Pollitt, C. 2012. *New Perspectives on Public Services: Place and Technology.* Oxford, Oxford University Press.

Tammaru, T., Leetmaa, K, Silm, S. and Ahas, R. 2009. Temporal and spatial dynamics of the new residential areas around Tallinn. *European Planning Studies*, 17(3): 423–39.

Terk, E. 1999. *Erastamine Eestis: ideoloogia, läbiviimine, tulemused.* Tallinn: Eesti Tuleviku-uuringute Instituut.

Thorhallsson, B. and Kattel, R. 2012. neo-liberal small states and economic crisis: lessons for democratic corporatism. *Journal of Baltic Studies* [Online], 1–21, iFirst. Available at: http://uni.hi.is/baldurt/files/2012/10/Neo-liberal-small-states-Estonia-Iceland-by-Thorhallsson-Kattel.pdf [accessed: November 14, 2012].

Tiits, M. 2007. Technology-intensive FDI and economic development in a small country—the case of Estonia. *Trames: Journal of the Humanities and Social Sciences*, 11 (61/56)(3): 324–42.

Varblane, U. 2010. Foreign Direct Investment in Estonia, in *Foreign Investors in Estonia: 15 Case Studies*, edited by D. Scott. Tallinn: Tallinna Raamatutrükikoda.

Vida, I. and A. Fairhurst. 1998. International expansion of retail firms: a theoretical approach for future investigations. *Journal of Retailing and Consumer Services*, 5(3): 143–51.

Wrigley, N. and Lowe, M. 2002. *Reading Retail*. New York. Arnold, London: Oxford University Press.

Newspapers, Reports, and Interviews

Alts. 2012. *Pilt: Tartu linnaosad.png*. *Wikipedia*, April 21, 2012 [Online]. Available at: http://et.wikipedia.org/wiki/Pilt:Tartu_linnaosad.PNG [accessed: December 13, 2012].

Baltic News Service. 2001. Kaubanduspindade arendamine liigub Tallinnast väljapoole. *Eesti Päevaleht*, February 16, 2001 [Online]. Available at: http://www.epl.ee/artikkel/147025 [accessed: March 28, 2011].

Bank of Estonia. 2011. International Investment Position [Online]. Available at: http://statistika.eestipank.ee/?lng=et#listMenu/814/treeMenu/MAKSEBIL_JA_INVPOS/146 [accessed: November 15, 2011].

Centre of Registers and Information Systems. 2012. *Annual Reports of Prisma Peremarket Ltd., Rimi Eesti Food Ltd., Selver Ltd., Maxima Ltd., Eesti Tarbijateühistute Keskühistu, Comarket Ltd. 2002–2010* [Online]. Available at: https://ariregister.rik.ee/ [accessed: February 15, 2012].

EIER. 2003. *Eesti toidukaupade positsioon siseturul. Kodumaiste ja importtkaupade osakaal Eesti kaupluse käibes ja sortimendis ning hinnad.* Estonian Institute of Economic Research, Tallinn.

EIER. 2010. *Eesti toidukaupade positsioon siseturul. Kodumaiste ja importtkaupade osakaal Eesti kaupluse käibes ja sortimendis ning hinnad.* Estonian Institute of Economic Research, Tallinn.

EMOR. 2010. *Jaekaubanduse seire 2010. September–oktoober 2010.* TNS EMOR, Eesti Toiduainetööstuse Liit [Online]. Available at: http://www.agri.ee/public/juurkataloog/UURINGUD/turuarendustoetus_uuringud/uuring_turuarendustoetus_2010_jaekaubandus-seire.pdf [accessed: December 14, 2011].

EPL. 2001. Baltic Foodi ajas lõhki iha suure tüki järele. *Eesti Päevaleht*, November 12, 2011 [Online]. Available at: http://www.epl.ee/artikkel/181011 [accessed: December 4, 2011].

EPLO. 2003. Linn ei anna Säästumarketi ehitamiseks luba. *Eesti Päevaleht Online*, November 18, 2003 [Online]. Available at: http://www.epl.ee/artikkel/250696 [accessed: April 30, 2011].

Erala, S. 2009a. Rimi tõi müügile omanimelised leivad. *Postimees*, September 21, 2009 [Online]. Available at: http://naine.postimees.ee/?id=166316&redir= [accessed: August2, 2011].

Erala, S. 2009b. Rimi kaubavalikusse lisanduvad ajutiselt reisid. *Postimees*, September 29, 2009 [Online]. Available at: http://www.tarbija24.ee/?id=169452 [accessed: April 4, 2011].

Erala, S. 2010. Rimi jätkab suunamuutust ja hakkab müüma Eesti talu piimatooteid. *Postimees*, September 9, 2010 [Online]. Available at: http://www.tarbija24.ee/?id=310795 [accessed: May 25, 2011].

Erilaid, E. 1999. Eesti Spar müüakse Rootsi kaupmeestele. *Postimees*, June 2, 1999 [Online]. Available at: http://arhiiv2.postimees.ee:8080/leht/99/06/02/majandus.shtm [accessed: April 22, 2011].

ETA. 2002. Linnakohus kuulutas välja Baltic Foodi Eesti pankroti. *Eesti Päevaleht*, 24 January 2002 [Online]. Available at: http://www.epl.ee/artikkel/189779 [accessed: September 14, 2011].

Gnadenteich, U. 2009. Nõmme turg liideti Tallinna abipaketiga. *Tallinna Postimees.ee*, 2 April 2009 [Online]. Available at: http://www.tallinnapostimees.ee/?id=102750 [accessed: November 22, 2011].

Hankewitz, D.G. 2010. Rimi: müüme vaid imporditud värsket sea- ja veiseliha. *Postimees*, April 9, 2010 [Online]. Available at: http://www.e24.ee/?id=247770 [accessed: May 23, 2011].

Hint, G. 2002. Uus toiduaineseadus suretab väikepoed välja. *Eesti Päevaleht*, July 17, 2002 [Online]. Available at: http://www.epl.ee/artikkel/210050 [accessed: 5 November 2011].

Josing, M. 2007. Mis toimub hindadega? *Eesti Päevaleht*, September 27, 2007 [Online]. Available: at http://www.epl.ee/artikkel/401630 [accessed: April 24, 2011].

Kallas, R. 2011. Omatooted labidast lihani? *Delfi Majandus*, March 17, 2011 [Online]. Available at: http://majandus.delfi.ee/news/tarbija/margus-venelaine-omatooted-labidast-lihani.d?id=42268853 [accessed: April 9, 2011].

Karotamm, L. 2008. Maxima: eikusagilt kõrgustes. *Postimees*, March 21, 2008 [Online]. Available at: http://www.postimees.ee/150508/esileht/majandus/318902.php [accessed: September 20, 2011].

Kass, M. 2010. Seeder: Rimi süüdistused Eesti liha kvaliteedi kohta on alusetud. *Postimees*, April 9, 2010 [Online]. Available at: http://www.tarbija24.ee/247634/seeder-rimi-suudistused-eesti-liha-kvaliteedi-kohta-on-alusetud/ [accessed: May 24, 2011].

Kivistik, L. 2002. Tallinna kaubanduspindu ärvardab ülepakkumine. *Eesti Päevaleht*, May 22, 2002 [Online]. Available at: http://www.epl.ee/artikkel/204980 [accessed: March 28, 2011].

Koovit, K. 2009. Kaupluseketid valmistuvad turu kukkumiseks. *Postimees*, February 17, 2009 [Online]. Available at: http://www.postimees.ee/?id=83240 [accessed: March 28, 2011].

Lohk, M. 2007. Tarbijakaitseamet algatas Maxima suhtes 17 väärteomenetlust. *Postimees*, November 1, 2007 [Online]. Available at: http://www.tarbija24.ee/240308/esileht/olulised_teemad/tarbija24/kasu/292975.php [accessed: April 13, 2011].

Lohk, M. 2010a. Põllumehed kutsuvad üles Rimi importliha boikoteerima. *Postimees*, April 9, 2010 [Online]. Available at: http://www.tarbija24.ee/?id=247623 [accessed: May 25, 2011].

Lohk, M. 2010b. Rimi tõi müügile Rootsi emafirma ökosarja. *Postimees*, January 20, 2010 [Online]. Available at: http://www.tarbija24.ee/?id=214119 [accessed: May 26, 2011].

Mallene, L. 2011. Ma olin Maxima ori. *Eesti Ekspress*, September 1, 2011. Available at: http://www.ekspress.ee/news/paevauudised/eestiuudised/ma-olin-maxima-ori.d?id=56864916 [accessed: January 7, 2012].

Männiste, A. 2006. Maxima pakub odavamaid hindu. *Postimees*, September 21, 2006 [Online]. Available at: http://www.tarbija24.ee/280906/esileht/olulised_teemad/tarbija24/toit/219124.php [accessed: February 26, 2011].

Masing, K. 2007. Selver ja Rimi ähvardavad alkoholipiirangu korral poed kell 20 sulgeda. *Eesti Päevaleht*, July 14, 2007 [Online]. Available at: http://www.epl.ee/artikkel/389769 [accessed: June 28, 2011].

Merilai, M. 2011. *Interview*. Estonian Traders Association. Transcription: June 14, 2011.

Mmh. 2008. *Pilt: Tallinna linnaosad.png*. *Wikipedia*, September 14, 2008 [Online]. Available at: http://et.wikipedia.org/wiki/Pilt:Tallinna_linnaosad.png [accessed: December 13, 2012].

Murde, R. 1995. Turud pärsivad kaupluste arengut. *Eesti Päevaleht*, November 27, 1995 [Online]. Available at: http://www.epl.ee/news/majandus/turud-parsivad-kaupluste-arengut.d?id=50721472 [accessed: May 11, 2011].

Neudorf, R. 2002. Väikesed kauplused sulgevad uksed. *Eesti Päevaleht*, April 24, 2002 [Online]. Available at: http://www.epl.ee/artikkel/202353 [accessed: July 3, 2011].

Niitra, S. 2010. Kõige tervisesõbralikum töökoht on Selver. *Postimees*, November 25, 2010 [Online]. Available at: http://www.tarbija24.ee/?id=347609 [accessed: September 19, 2011].

OECD. 2000. *OECD Regional Economic Assessment: The Baltic States*. Organization for Economic Cooperation and Development.

Palts, M. 2011. *Interview*. Estonian Chamber of Commerce and Industry. Transcription: May 4, 2011.

Paves, K. 2010. Minister tahab karmistada alkoholimüügi korda. *Postimees*, June 2010 12, [Online]. Available at: http://www.e24.ee/?id=275404 [accessed: October 30, 2011].

Peensoo, P. 2003. Lilleküla kerkib Citymarket. *Eesti Päevaleht*, May 6, 2003 [Online]. Available at: http://www.epl.ee/artikkel/234765 [accessed: March 24, 2011].

Pirn, T. 1997. ÜHINEMINE: Kaubahall ühines 27 riigis esindatud jaekaupluste ketiga Spar. *Eesti Päevaleht*, December 13, 1997 [Online]. Available at: http://www.epl.ee/artikkel/28949 [accessed: June 5, 2011].

PM Online. 2002. Suurpoed ei taha väikeste Eesti firmade kaupa. *Postimees*, July 15, 2002 [Online]. Available at: http://vana.www.postimees.ee/index.html?number=919&op.=lugu&id=64512 [accessed: June 9, 2011].

Potisepp, S. 2010. Sirje Potisepp: Eesti liha toetuseks. *Postimees*, March 30, 2010 [Online]. Available at: http://www.e24.ee/?id=243620 [accessed: February 10, 2011].

Potisepp, S. 2011. *Interview*. Association of Estonian Food Industry. Transcription: July 1, 2011.

Reimer, A. 2009. Väikepoed sünnivad taas—esinduslikes kaubakeskustes. *Eesti Päevaleht*, March 18, 2009 [Online]. Available at: http://www.epl.ee/artikkel/462385 [accessed: September 22, 2011].

Reinap, A. 2003. Kommentaar: Pankade käibemaks. *Postimees*, December 30, 2003 [Online]. Available at: http://wwx.postimees.ee/311203/esileht/majandus/122832.php [accessed: June 26, 2011].

Riikoja, H. 2006. Kaupluseketid võtavad tänavu ette laienemise. *Postimees*, January 10, 2006 [Online]. Available at: http://www.postimees.ee/110106/esileht/majandus/188325.php [accessed: April 2, 2011].

Roonemaa, H. 2003. Väikesed toidupoed asuvad Tallinna turgu vallutama. *Postimees*, September 1, 2003 [Online]. Available at: http://vana.www.postimees.ee/index.html?number=919&op.=lugu&id=113285 [accessed: May 22, 2011].

Salu, M. 2008. Selveri juht Iivi Saar: valmis hakklihakastet ja kalleid asju ostetakse vähem. *Eesti Päevaleht*, September 24, 2008 [Online]. Available at: http://www.epl.ee/artikkel/442610 [accessed: December 2, 2011].

SEB. 2011. *Annual Reports 2002–2010.* SEB Bank.

Seaver, U. 2009. Piritalased protestivad Maxima poe vastu. *Postimees*, January 14, 2009 [Online]. Available at: http://www.tallinnapostimees.ee/?id=69141 [accessed: February 3, 2011].

Sikk, R. 2009. Parim toidupood on Selver, teistest viletsam Comarket. *Eesti Päevaleht*, March 28, 2009 [Online]. Available at: http://www.epl.ee/artikkel/463571 [accessed: April 26, 2011].

Sokk, E. 2011. *Interview*. Estonian Ministry of Agriculture. Transcription: July 13, 2011.

Statistics Estonia. 2011. *Statistical database. Economy* [Online]. Available at: http://www.stat.ee/economy [accessed: October 20, 2011].

Sõrm, M. 2011. *Interview*. Estonian Ministry of Economic Affairs and Communications. Transcription: July 18, 2011.

Sõrmus, R. 2011. *Interview.* Estonian Chamber of Agriculture and Commerce. Transcription: May 26, 2011.

Swedbank. 2011. *Annual Reports 2002–2010.* Swedbank Estonia.

Tallinn Official Website. 2012. *Tallinna elanike arvud 2010* [Online]. Available at: http://www.tallinn.ee/est/Tallinna-elanike-arvud-2010 [accessed: December 15, 2012].

Tammela, A. 2009. Maxima korraldab laada. *Postimees*, November 20, 2009 [Online]. Available at: http://www.24tundi.ee/?id=190654 [accessed: October 2, 2011].

Tartu Linnavalitsus. 2012. *Statistiline* Ülevaade. *Tartu 2011* [Online]. Available at: http://www.tartu.ee/data/tartu_stat_aastaraamat_2011_veeb.pdf [accessed: December 10, 2012].

Taube, A. 2005. Jaekaubanduses saab kord majja. *Tartu Postimees*, June 13, 2005 [Online]. Available at: http://tartu.postimees.ee/150605/tartu_postimees/168663.php [accessed: August 14, 2011].

Tigasson, K.R. 2002. Eesti turgude käive langeb, kaupluste oma tõuseb. *Postimees*, September 3, 2002 [Online]. Available at: http://vana.www.postimees.ee/index.html?number=919&op.=lugu&id=70138 [accessed: March 15, 2011].

Tomak, T. 1997a. Estiko laiendab SPAR-ketti. *Postimees*, August 15, 1997 [Online]. Available at: http://arhiiv2.postimees.ee:8080/leht/97/08/15/majandus.htm [accessed: September 28, 2011].

Tomak, T. 1997b. Kaubanduse ketistumine hoogustub. *Postimees*, September 25, 1997 [Online]. Available at: http://arhiiv2.postimees.ee:8080/leht/97/09/25/majandus.htm [accessed: October 1, 2011].

Villak, H. 2010. Kaupluseketi omatoode muutub ostja jaoks järjest tähtsamaks. *Postimees*, May 5, 2010 [Online]. Available at: http://www.e24.ee/?id=258757 [accessed: February 18, 2011].

Voog, A. 2011. *Muutused tarbijakäitumises aastatel 1991–2011*. TNS EMOR, October 26, 2011 [Online]. Available at: http://www.emor.ee/muutused-tarbijakaitumises-aastatel-19912011/ [accessed: February 5, 2012].

Chapter 7
The Role of Local and Foreign Companies in the Development of Non-Metropolitan Rural Areas in Poland

Magdalena Dej

Introduction

For several years now, a discussion has been held in the international research literature on the importance of foreign capital in local and regional development. Various advantages and disadvantages are indicated resulting from the location of foreign companies in a specific area. Even though foreign researchers frequently emphasize a number of disadvantages deriving from the impact of external entities and point to relatively little integration of foreign investors with the local environment and little use of local resources, some Polish researchers who undertake this task indicate a relatively beneficial impact on local development, which depends on a variety of features of an investment location. The purpose of the chapter is to make a contribution to the discussion, which has been going on for a while and to show, using the example of non-metropolitan rural areas of Poland, the impact of large foreign economic entities[1] on the local community in comparison to the impact of Polish companies.

In the chapter, detailed attention is drawn to:

- the analysis of the distribution and characteristics of large foreign companies in non-metropolitan rural areas of Poland;
- the presentation, in relation to the entire country, of the development level of communes (townships) in non-metropolitan rural areas featuring large foreign companies in comparison to the development level of analogous communes featuring Polish companies and communes without any large economic entities.
- the determination, using the example of twelve large enterprises, of the impact of such entities on the local community depending on the origin of capital (foreign entities *versus* local Polish companies).

1 Large companies are understood as enterprises that employ at least 200 persons.

Foreign and Local Companies, Embeddedness and Impact on the Local Community

There exist different manners of depicting the embeddedness of an economic operation (Oinas 1997, 1998); however, all take into account mutual relations between the analyzed entity and the local community. On this basis, Micek, Działek and Górecki (2010: 15) assumed that (…)

> the embeddedness of an enterprise consists of its participation in a wide spectrum of social and economic relationships with the environment (local community). Principal relationships constituting the local embeddedness of an enterprise include ties with suppliers and customers, contacts with local authorities and institutions, actions taken by the enterprise for the benefit of the local community, and employee relationships (including managerial personnel) with the local community.

Therefore, embeddedness is to be perceived as a broader term in relation to the impact of companies on the local community. The impact of companies takes into account unilateral actions—company activities targeted towards the local community—whereas embeddedness takes into account bilateral actions. Studies on the embeddedness of foreign economic entities in Central and Eastern Europe including Poland have been conducted in large numbers. At the same time, it is necessary to pay attention to the differences in the conclusions reached by various researchers. In the works of Polish researchers, strong ties between the analyzed entities and local communities can be inferred (B. Domański 2004, Stryjakiewicz 2004, B. Domański and Gwosdz 2005). For example, Bolesław Domański (2004: 21) argues:

In general, the embeddedness of large foreign manufacturers is quite significant in Poland, which has had a positive impact on the competitiveness of entire geographic regions and individual locations alike, primarily via an improvement in the prospects of local enterprises and an increase in the qualifications of the local labor force.

Such entities, as a result of paying local taxes, complying with local environmental provisions and creating good working conditions, may be perceived as good local citizens (B. Domański 2004: 16). At the same time, contradictory theses appear in the works of foreign researchers. In 1994, Grabher, in reference to foreign investors doing business in Eastern Germany, used the metaphor of *cathedrals in the desert*, which was supposed to show the nature of their relationships with local communities, manifested by very weak relations with the local economy, and thence slight impact on regional development. This term was subsequently used by other researchers including J. Hardy (1998) who reached his conclusion concerning the insignificant integration of foreign companies with the local community on the basis of selected enterprises operating in the Wrocław area. On the other hand, when describing relationships between foreign investors and the local economy in Central and Eastern Europe, Burbach, Nunez and Kagarlitsky (1997) used the term Kuwaitization, which is understood as the creation, by international corporations, of so-called *strong points* with colonial features. Such points are isolated from their surrounding territory and their characteristic feature is deindustrialization.

Bolesław Domański (2001) draws attention to irregularities in the manner of perceiving the impact of foreign investors by some researchers from the broadly understood Western circle of academics when discussing methodological and ideological deformations in modern studies on economic phenomena in Central and Eastern Europe. Bolesław Domański stresses that it is a mistake to underscore the role of cheap labor as the principal factor attracting foreign investors as a theoretical concept. Empirical studies clearly show that many foreign entities are oriented towards the acquisition of the target country's domestic market. Additionally, in contrast to what has been written by some foreign researchers,

> work conditions at the plants of large enterprises located in Poland have been shown to be—both by experts and trade union activists—similar as those in analogous plants in the "old" European Union [...]. Managers commonly emphasize the significance of work quality enabling quick achievement of production quality comparable with Western Europe and increases in efficiency. Studies suggest that the basic benefit for companies investing in Poland with the labor force in mind is the possibility of selecting the best employees thanks to advantages over local competitors in local labor markets. (Domański 2001: 29)

Another mistake, according to Bolesław Domański, is the underestimation of local factors when analyzing the causes of regional and local development. Even in areas with an exceptionally high concentration of large foreign companies (Tarnowo Podgórne, Niepołomice, Gliwice), local factors play a very important role. Even more so, in areas devoid of external development factors, local factors play a decisive role in development.

Taking into account solely foreign investors, Stryjakiewicz (2004) notices that their impact on the local community results from their adopted strategy of operation. Furthermore, he emphasizes that:

> It is understandable that in the strategies of a global concern, regional and local dimensions have secondary significance [...] and that such strategies take into account the expectations and preferences of cities and entire regions to only a slight degree. However, past experience [...] has shown that strong embeddedness of a foreign enterprise in its local community may be the source of benefits or even development impulses for both the company and its corresponding geographic region. (Stryjakiewicz 2004, 80)

Research Methods

The analysis was conducted on two levels of specification. First, comparative studies covered all communes (townships) located in non-metropolitan rural areas featuring large enterprises operating with and without foreign capital. Second, case studies were conducted for 12 enterprises and communes selected using specific criteria.

The study area was defined as rural communes, excluding rural communes whose seats are located in cities and those found outside the borders of officially designated metropolitan areas in Poland (Smętkowski, Jałowiecki and Gorzelak 2008). nMRAs[2] (non-metropolitan rural areas) determined in this manner corresponds to types of *remote–rural areas* and the majority of *easily accessible areas* according to the division applied by Keeble and Tyler (1995) and Anderson, Tyler and McCallion. (2005).

For the purpose of identifying large enterprises, a query was conducted using three telephone and address databases containing information on employment volume at companies. Each time, the inclusion of a given non-metropolitan commune in the analysis was determined by the presence, within its borders, of an enterprise having at least 200 employees. As a result of the conducted search query, it was possible to identify 224 such companies operating in 171 communes. In this group, companies with a majority share of foreign capital constituted 21 percent of all large companies in nMRAs (47 entities). Subsequently, indices describing the social and economic situation, as well as the situation within the scope of spatial management of communes with large companies, with differentiation into entities with Polish capital and a majority share of foreign capital as well as remaining communes without large enterprises were compared. Comparisons were conducted with the use of single-factor variance analysis—ANOVA.[3] Values of partial and synthetic indices were calculated using the Perkal method. In the literature, the Perkal method is known as the Z-score method (Smith 1972). This index, using standardized values of indices for each of the analyzed units, calculates their arithmetic average.

For the purpose of analyzing the impact of large enterprises on local communities depending on the origin of capital (foreign capital vs. local Polish companies) in the period between November 2009 and April 2010, field studies were conducted at 12 selected enterprises. Research studies conducted at companies were supplemented

2 nMRAs include 54 percent of all communes in Poland and are inhabited by approx. 8.3 million people.

3 Single-factor variance analysis (ANOVA) is a method used to compare values of average indices among populations. Furthermore, it provides information whether the differences are statistically significant. In the case of comparing two average values, ANOVA gives the same results as the t test for independent samples (in the case of comparing two various groups of cases/observations) or the t test for dependent samples (when two variables in the same groups of cases/observations are compared).

with interviews with managing personnel as well as a questionnaire concerning numerical data on each given company. The questionnaire was not completed by two companies. In this manner, information about the size and the structure of employment was obtained, along with changes in this respect in the last couple of years. Other data obtained included workplace quality, wages and salaries, and place of residence of employees with respect to the positions held by them, as well as ties with local economic entities.

Questionnaire studies were conducted among the inhabitants of the analyzed communes. The purpose of the questionnaire was to identify the inhabitants' opinions on each large enterprise as an employer. Of the 4,900 questionnaires distributed, 52 percent were completed (2,533).

Distribution of Large Companies in nMRAs vs. Origin of Capital

Out of the 224 identified enterprises, the presence of foreign capital (total or majority) was ascertained in the case of 47. German capital was identified at the largest number of companies (14). French capital was identified at seven companies, Austrian capital at six, American and Belgian capital at three, as well as Danish, Dutch, British and Finnish capital at two companies each. Furthermore, Spanish, Japanese, Norwegian, Slovak, Italian, Czech and Chinese capital investments were also identified. The geographic distribution of companies has been presented in Dej (2009).

From the point of view of impact on the local labor market, *greenfield* investments, which generate new jobs, have particular significance. Out of 47 large companies with foreign capital operating in nMRAs, a definite majority, that is as many as 35 entities are *greenfield* type companies. Approximately 13,000 jobs were created by these 35 companies. On the other hand, 6,900 employees work at the remaining 12 plants taken over by foreign investors. Takeovers of Polish companies occurred most frequently (4 times) in the Kujawy and Pomorze provinces. Two companies of this type were also identified in the following five provinces: Małopolskie, Podkarpackie, Śląskie, Lubelskie, and Łódzkie.

The degree of technological sophistication of an industrial company is an indicator of its level of innovation, which is associated with its level of research and development. One way to measure this is to relate R&D expenditures and the value of goods sold or value added (Nauka i technika 2011). The 2009 Eurostat classification system identifies four categories of R&D activity: high-tech, medium high-tech, medium low-tech, and low-tech.

In the case of 139 large Polish companies located in nMRAs, a definite majority (75.5 percent) are entities with the lowest level of technological advancement, whereas among entities with a majority share of foreign capital, the share of low-tech companies does not exceed 50 percent (Table 7.1). In the second group, most companies are classified as medium-low technology. Moreover, the second group includes companies that are very advanced technologically. All large companies with the highest level of technological advancement located in Polish nMRAs are entities with foreign capital. This includes companies such as Orion and Sharp in Łysomice near Toruń, Kimberly-Clark in the strongly industrialized Klucze commune near Olkusz, and ZF Altana Pharma (currently Nycomed Pharma) in Łyszkowice near the city of Łódź.

A comparison of the degree of technological sophistication of large industrial companies in nMRAs, regardless of the source of capital, with that of all companies in Poland with more than nine employees (Table 7.1) does not unequivocally show that companies in nMRAs are less innovative. The only discernible difference applies to large low-tech companies, which are more numerous—more than 14 percentage points more—in nMRAs than the national average for companies with more than nine employees.

Table 7.1 Level of advancement of large industrial companies in nMRAs depending on the origin of capital versus the general level of technological advancement of companies in Poland

Level of technological advancement of industrial companies	Large industrial companies in nMRAs						Companies in Poland (>9 employees)*
	Polish		Foreign		Total		
	Number of companies	Share (%)	Number of companies	Share (%)	Number of companies	Share (%)	Share (%)
Low technology	105	75.54	21	48.84	126	69.23	55.08
Medium-low technology	27	19.42	12	27.91	39	21.43	36.82
Medium-high technology	7	5.04	6	13.95	13	7.14	6.33
High technology	0	0.00	4	9.30	4	2.20	1.77
Total	115	100.00	43	100.00	182	100.00	100.00

Note: * calculations based on Bank of Local Data (2010).
Source: own studies.

Level of Development of Communes Depending on the Presence of a Large Company and Origin of Capital

For the purpose of determining differences in the level of development of communes depending on the presence of a large enterprise and the origin of capital (Polish companies vs. foreign companies), analysis was performed at the level of all companies in nMRAs. Taking into account a number of indices classified into three groups—social, economic and spatial management—differences in the values of such indices are shown for communes with large companies versus communes without such entities, with the use of single-factor variance analysis ANOVA. Subsequently, an analogous analysis was conducted for a group of communes with large companies, comparing the level of development of communes depending on the type of invested capital—companies with a majority share of foreign capital vs. Polish companies. Data obtained from Poland's Central Statistical Office was used for the analysis in the chapter. The analysis was conducted on the basis of partial indices (shown in Table 7.2, Table 7.3 and Table 7.4), as well as comprehensive indices (social, economic and spatial management) constructed on their basis with the use of the Z-score method.

Table 7.2 Comparison of average social index values for nMRA communes with large companies depending on the origin of capital (single-factor variance analysis ANOVA)

Index (abbreviated name)	Communes with large companies	Number of communes	Average	Standard deviation
Age pyramid index	Polish	130	1.54	0.38
	Foreign	41	1.65	0.39
Percentage of women	Polish	130	101.49	3.49
	Foreign	41	102.18	2.41
Internal migration balance	Polish	130	2.47	11.35
	Foreign	41	5.27	13.47
Foreign migration balance	Polish	130	–0.40	1.19
	Foreign	41	–0.23	0.59
Percentage of children in preschools	Polish	130	32.24	23.45
	Foreign	41	35.15	22.34
Percentage of people with a higher education	Polish	130	4.66	1.47
	Foreign	41	5.17	2.50
Percentage of apartments with bathrooms	Polish	130	77.84	11.57
	Foreign	41	79.52	9.85
Number of stores per 1,000 inhabitants	Polish	130	8.00	2.55
	Foreign	41	8.97	2.27

Source: own studies.

Table 7.3 Comparison of average economic index values for nMRA communes with large companies depending on the origin of capital (single-factor variance analysis ANOVA)

Index (abbreviated name)	Communes with large companies	Number of communes	Average	Standard deviation
Unemployment rate	Polish	130	8.34	3.99
	Foreign	41	7.65	2.90
Number of people commuting to work	Polish	130	0.88	1.93
	Foreign	41	3.12	14.90
Percentage of people working in agriculture	Polish	130	5.25	9.89
	Foreign	41	5.02	6.02
Economic entities	Polish	130	63.55	26.34
	Foreign	41	66.07	23.05
Commune's tax revenue	Polish	130	883.36	507.82
	Foreign	41	1805.78	5176.87
Commune's share of national personal income tax revenue	Polish	130	313.97	158.10
	Foreign	41	332.96	196.13
Commune's share of national corporate tax revenue	Polish	130	16.13	67.01
	Foreign	41	109.65	531.34

Source: own studies.

Table 7.4 Comparison of average spatial management index values for nMRA communes with large companies depending on the origin of capital (single-factor variance analysis ANOVA)

Index (abbreviated name)	Communes with large companies	Number of communes	Average	Standard deviation
Percentage of arable land	Polish	130	75.19	13.58
	Foreign	41	77.52	12.32
Percentage of other types of land	Polish	130	16.34	8.52
	Foreign	41	18.24	9.74
Entities using the water supply network	Polish	130	75.21	23.27
	Foreign	41	82.76	15.82
Entities using the sewerage system	Polish	130	25.78	19.41
	Foreign	41	33.90	22.65
Entities using municipal natural gas	Polish	130	19.46	25.53
	Foreign	41	17.66	25.07
New apartments per 1,000 inhabitants	Polish	130	2.97	2.57
	Foreign	41	3.80	4.77

Source: own studies.

The conducted analysis makes it possible to ascertain that communes with large companies found in nMRAs clearly differ from communes without such enterprises, which is marked both at the level of partial and comprehensive indices (Figure 7.1).

However, the particularly important fact is that the analysis allows for confirming that the level of development of communes with foreign companies is much higher than that of communes with Polish companies, and this dependence is associated with local social and economic issues as well as spatial management issues. In the case of every comprehensive index described herein, values obtained for communes with large foreign companies are much higher than those for communes with Polish companies, and even higher than those for communes without any large companies (Figure 7.1). However, this is not tantamount to finding an exceptionally positive impact of large foreign companies on local communities in Poland, as the high level of development of communes with large foreign companics may constitute a factor attracting other external investors, including foreign investors. An analysis of the distribution of large companies in nMRAs indicates that external investors very frequently select less peripheral locations in more urbanized and industrialized areas. Such areas are naturally more developed than areas that are less urbanized. In order to provide a clear answer to the question concerning the actual impact of foreign companies on local communities relative to the impact produced by Polish companies, it is necessary to discuss the results of field studies conducted at twelve companies and corresponding communes.

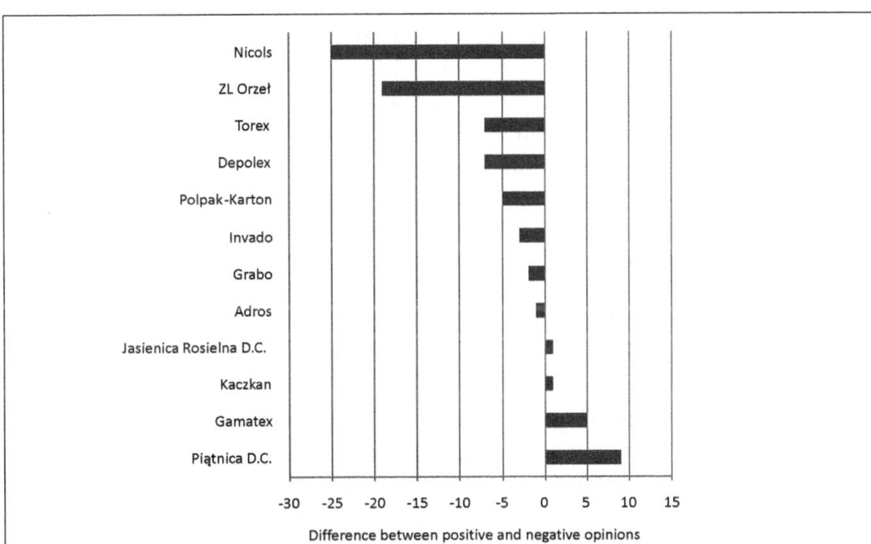

Figure 7.1 Variation of development level of communes in nMRAs depending on the presence of large companies and origin of capital (single-factor variance analysis ANOVA)

Source: own studies.

Results of Field Studies

Twelve companies and communes constituting their local environment were selected for the research study. The following criteria were used to evaluate the companies and communes: size of employment, profile of operation, origin of capital, period of establishment, key economic features of a commune. An attempt was made to select companies in a manner that they represent various categories within the scope of the above listed criteria, corresponding to an analogous structure for all large companies located in nMRAs. For 2008, three fourths of all the selected companies were entities with Polish capital—all were local companies established by local residents. The remaining one fourth were companies with majority foreign capital—German and Belgian—including Depolex, Nicols and Torex. Detailed information on the selected companies including their location, profile of operation, size of employment, year of establishment, and origin of capital is provided in Table 7.5.

The companies selected for the research study are located in 10 out of 16 Polish provinces. In spite of the fact that all the analyzed entities function in nMRAs, they differ in terms of their actual distance to municipal centers. Geographic location was a factor in the company selection process. Suburban companies located in communes adjoining poviat cities (ZL Orzeł near Jelenia Góra, Piątnica D.C. near Łomża, Gamatex near Nowy Targ, Depolex near Leszno) were selected, as were companies close enough to cities that daily commuting is possible (Torex, relatively near Łódź; Nicols relatively near Lublin). Some of the companies also selected were found at even more peripheral locations (Kaczkan, Polpak-Karton, Adros, Invado, Grabo, Jasienica Rosielna D.C.).

The study confirmed the immense significance of the origin of capital for differentiating the impact of large companies located in nMRAs. The research results show that the origin of capital affects the impact of companies on nMRAs in terms of the types of jobs offered to local employees, working conditions, stability of employment, and wage levels. It also affects local business conditions via ties with local suppliers of goods and services.

Types of Jobs Available to Local Workers

Management and engineering jobs tend to be filled in a manner that is different from that applicable to manufacturing and secretarial positions. The latter tend to be filled in a similar manner regardless of the type of company. The former can be separated into two types. The first group of companies satisfies all its personnel needs, including that for highly qualified employees, in the local job market (commune or poviat). These are companies, which are strongly tied to their local community by a long presence in that given community (ZL Orzeł, Jasienica Rosielna D.C.). In some cases, the connection results from the owners being members of the community (Invado, Polpak-Karton). On the other hand,

Table 7.5 General characteristics of enterprises selected for field study

Company	Commune	Poviat/ Province	Profile of operation	Employment	Year of establishment	Majority capital
Adros	Dobrzyca	Pleszew/ Wielkopolska	Meat processing	420	2000	Polish
Depolex	Lipno	Leszno / Wielkopolska	Production of upholstery	500	1989	German
Gamatex	Czarny Dunajec	Nowy Targ / Małopolska	Production of clothing	200	1975	Polish
Grabo	Dydnia	Brzozów / Podkarpackie	Wood processing	245	1991	Polish
Invado	Ciasna	Lubliniec / Opole	Production of doors	250	1990	Polish
Kaczkan	Małdyty	Ostróda / Warmia and Mazury	Wood processing	263	1987	Polish
Nicols	Trawniki	Świdnica / Lublin	Production of cleaning agents	200	1994	Belgian
Jasienica Rosielna D.C.	Jasienica Rosielna	Brzozów / Podkarpackie	Production of milk and other dairy products	206	1927	Polish
Piątnica Dairy Cooperatives (D.C.)	Piątnica	Łomża / Podlasie	Production of milk and other dairy products	340	1979	Polish
Polpak-Karton	Bukowiec	Świecko / Kujawy and Pomorze	Production of cardboard packaging	245	1994	Polish
Torex	Piątek	Łęczyca / Łódź	Production of curtain rails	280	1992	German
ZL Orzeł	Mysłakowice	Jelenia Góra / Dolny Śląsk	Production of linen materials	685	1842	Polish

Source: own study.

the second group consists of companies relying on outside recruiting for higher level employees (Table 7.6). It is interesting to note that these are companies with foreign capital (Nicols, Torex, Depolex), located on the grounds of former collective farms (Adros, Depolex, Kaczkan). This situation is related to the higher personnel requirements of foreign companies, their lack of familiarity with the local labor market, and a lack of informal ties with the local community.

Table 7.6 Place of residence of employees vs. positions held by them

Types of positions	Majority of employees from:			
	Commune	Remaining part of the poviat	Provincial city	Other locations
production	Kaczkan, Nicols, Torex, Adros, Jasienica R. D.C., ZL Orzeł	Polpak-Karton, Depolex, Invado	–	–
clerical	Torex, Jasienica R. D.C., ZL Orzeł	Kaczkan, Polpak-Karton, Depolex, Invado	–	Adros
engineering	Jasienica R. D.C.	Kaczkan, Invado ZL Orzeł	Nicols	Torex, Depolex
management	Torex	Polpak-Karton, Invado ZL Orzeł	Kaczkan, Nicols	Adros

Source: own studies.

Quality of Jobs

It was also ascertained that type of capital diversifies the quality of jobs offered in terms of wages and stability of employment. First, it is necessary to draw attention to the general evaluation of companies as employers. As shown in Figure 7.2, all the examined companies with foreign capital were rated as employers by employees from the local community. The rating was rather poor compared to that for Polish companies with strong local relationships. The following companies were rated as best employers: Piątnica D.C., Jasienica Rosielna D.C., Gamatex, Invado and Polpak-Karton. In the case of foreign companies, most employees had a negative opinion of their employer. This was especially true in the case of the Nicols company, with the main complaints being low wages and instability of employment.

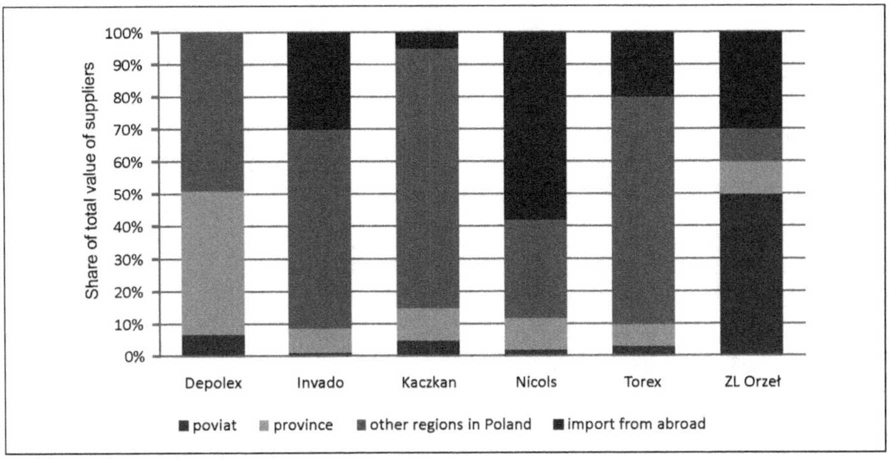

Figure 7.2 Opinions about companies as employers
Source: own studies.

Wage data were obtained from eight companies. Several companies did not provide wage data: Adros, Gamatex, Grabo and Piątnica D.C. Wage data are shown in Figure 7.3 and are compared with information on average wages in the poviats and provinces in Poland (Local Data Bank of the Central Statistical Office) where the examined companies are located. A significant analytical limitation appears in that companies provide wages for non-management employees only, whereas poviats and provinces provide average wage data for all employee groups. However, it is non-management employee wages that provide a more accurate picture of the economic impact of a company on its local community. Management salaries tend to inflate earnings averages, although most managers come from outside the local community, especially in the case of companies with majority foreign capital.

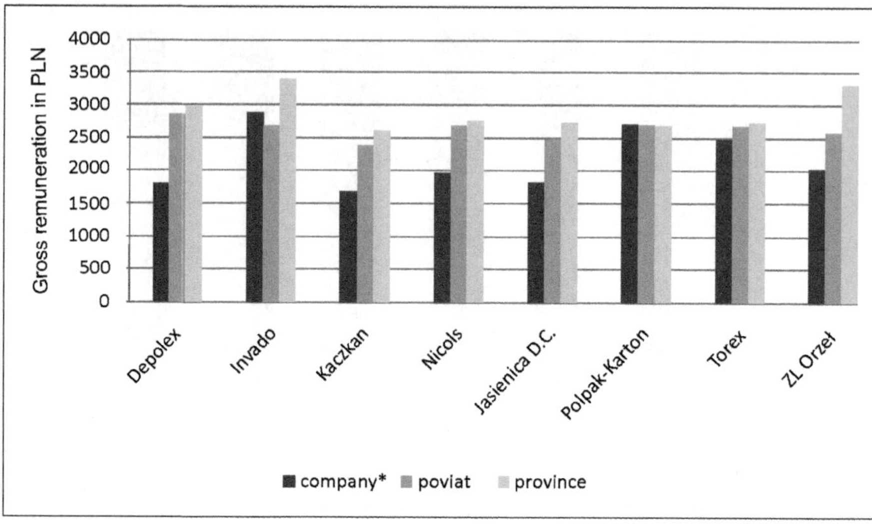

Figure 7.3 Gross monthly remuneration offered by the examined companies in comparison to remuneration averages in the local poviat and province

Note: * Excluding managerial personnel.

Source: own study on the basis of field studies and Local Data Bank of the Central Statistical Office.

The employees of Kaczkan definitely receive the lowest wages—PLN 1,700 gross pay—which is low even in comparison to Ostróda Poviat and the entire Warmia and Mazury Province. In this context, the exceptional unanimity of the respondents with respect to their negative evaluation of wages seems to be strongly justified. However, this is slightly contradicted by the generally good standing of the company Therefore, it seems that the wage level is determined primarily by peripheral location of the company and the conviction that on account of the absence of alternative workplaces in the local job market and a low probability of being employed outside one's place of residence, workers will decide to accept such work in spite of low pay.

Two other companies, Jasienica Rosielna D.C. and Depolex, offer only slightly higher wages—approx. PLN 1,800 gross pay. In both cases, wages were criticized by respondents. At the same time, it seems that due to evident social and economic differences across the country, which translate into costs of sustenance, inconveniences related to low wages were more perceptible for the employees of Depolex, which is located in the immediate vicinity of Leszno, than for employees from the Podkarpacie region of southern Poland. In the case of Depolex, low wages may be, to a certain extent, explained in the context of the very difficult financial situation of the company, which announced bankruptcy in 2010.

Nicols and ZL Orzeł were ranked next, with wages at approximately PLN 2,000 per month, while Torex offered almost PLN 2,500 per month. The highest wages are received by the employees of Polpak-Karton and Invado, PLN 2,700 and PLN 2,900 gross pay, respectively, which seems satisfactory compared to wages in their local communities. However, employees participating in the research survey were not satisfied with their wages.

Wages offered by Invado, a company with very good financial standing, correlate well with its positive image among local residents. The company is strongly tied to the Lubliniec job market, where it is the largest employer. The fact that it was established by members of the local community plays a certain role in this respect. The owners appear to understand the realities of life in this community. The same cannot be said of a company such as Nicols in Trawnica. The relative complexity of the production process and close proximity of a larger city make it difficult to maintain low wages, as does Kaczkan.

The foreign-owned Nicols company has a negative opinion among local residents, both regarding wages and the stability of employment. This is not consistent with the data made available by the company. In talks with the representatives of the company, opinions emerge concerning the minimal needs of rural residents. The company feels that such needs should be fully satisfied by the wages it offers. It was also possible to get the objective impression that the Nicols company moved to its current location in order to lower labor costs. At Polish companies, where the offered wages were even lower (Kaczkan, Jasienica Rosielna D.C.), it was possible to sense a much more subjective approach, which resulted from the fact that the owner and his employees live in the same town. The establishment of a local company by a local owner is not primarily determined by low labor costs.

The quality of jobs, apart from wages, is determined by several other factors, among which stability of employment has special significance. Employment stability is characterized by the percentage of employees employed on a full-time basis, percentage of outsourced employees, nature of employment contracts (fixed period of time or unlimited period of time), and HR policy (risk of dismissals).

Research has shown that the employees of large companies as well as members of the local community perceive the examined plants negatively in this respect. Lack of employment stability was the fourth—with respect to frequency—negative feature of workplace quality listed for these companies. At 8 out of 12 studied companies, negative opinions were prevalent in employee evaluations of employment stability. This is interesting in comparison to facts provided by employers. A clear majority of employers declared that at least 95 percent of jobs are full-time and the percentage of outsourced employees is of marginal importance. Furthermore, no company was planning any lay-offs over the next two years, even though lay-offs—in certain cases on a very large scale—had occurred over the five years preceding the research study.

At the same time, local residents working at Piątnica, Jasienica Rosielna, Gamatex and Kaczkan perceived the stability of employment as a factual advantage of their workplace, which is consistent with statements issued by their employers

(Figure 7.4). In the opinions of the inhabitants of the remaining communes, critical voices were dominant, especially in the case of the residents of Mysłakowice (ZL Orzeł) and Trawniki (Nicols). In the first case, the criticisms can be linked to the Orzeł company's financial instability, dismissals, and eventual bankruptcy. However, discrepancies in the evaluation of the Nicols company are rather surprising. Nicols management paints a picture of a stable company offering dependable jobs—98 percent full-time personnel, marginal percentage of outsourced employees, and job growth in the last couple of years. However, local residents do not confirm such claims, and instead indicate high rates of employee turnover and short and/or incomplete work contracts. Given the large number of such opinions among local residents, this version of events appears to be more plausible.

The image of job quality at companies is very complex. Research has shown that some employers enjoy a good reputation in their local community and others are perceived as a necessary evil by workers with few job alternatives. Undisputedly, the best studied employer was Piątnica D.C., which was rated as good in almost every aspect of its operation. The following three companies were also described as good employers—Jasienica Rosielna D.C., Gamatex, and Polpak-Karton. A less unanimous, but still quite positive picture is provided by the opinions of workers living near Invado, Kaczkan, and Torex. Finally, companies such as Depolex, Nicols, Adros, Grabo, and ZL Orzeł were perceived more negatively.

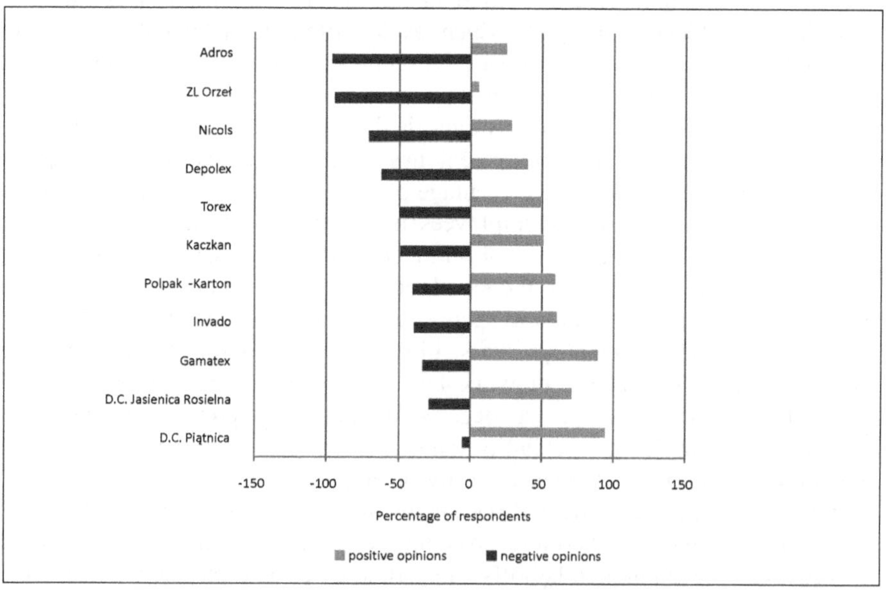

Figure 7.4 Opinions of local residents about the stability of employment at selected companies

Source: own study.

The evaluation of workplace quality, as perceived by local residents, has its confirmation in objective criteria. Job quality is definitely related to the financial standing of companies, as well as the relationship between owners, managers, and local residents. It seems that close ties resulting from simply living in a given geographic area combined with good economic standing of a company enhance job quality. Such observations may be inferred from the example of Piątnica D.C., Jasienica Rosielna D.C., Gamatex, Polpak-Karton, and Invado—companies with good financial standing, established or managed by persons from the local community. On the other hand, foreign companies with weak personal ties to the plant's home community, in particular Depolex and Nicols, attach less importance to workplace quality and tend to focus more on profitability of production.

Ties with the Local Community

Representatives from 6 out of 12 studied companies provided information about the approximate share of parts and other materials sourced locally and non-locally (home poviat, province, Poland, elsewhere)[4] (Figure 7.5).

First of all, there is very slight share of values of supplies from the nearest area of company community, excluding ZL Orzeł, where it amounts to 50 percent. At other companies, it is of marginal importance or is not relevant at all. It is characteristic that at Nicols, an entity with foreign capital, imports from abroad are of great significance—almost 60 percent of the total value of supplies. Therefore, this company is characterized by a very strong reliance on the local labor force and an almost complete lack of ties with local suppliers. Weak ties with local companies were also ascertained at Depolex which, in spite of being a company with foreign capital, relies solely on Polish suppliers.

Weak ties between large companies and local suppliers of parts and other materials are not a surprising phenomenon. In the era of global economic ties and progressing specialization of production, doing business with companies outside the local community is understandable and is not related to the peripheral location of a given company.

Apart from ties with parts and materials suppliers, economic ties may also exist with service providers who tend to be present in most communities. One measure of a service relationship is the number of local employees serving a large business enterprise. Service ties were analyzed for five companies performing a variety of services for large customers.

In this case, ties with the local community are stronger and are marked mainly by business with transportation companies. Each of the analyzed large companies uses the services of transportation firms from the local poviat. In the case of companies where weaker ties with the local community were ascertained, that

4 It was difficult to provide a specific share of production materials in the case of dairy cooperatives (Jasienica Rosielna D.C. and Piątnica D.C.) where milk supplies are of basic importance. Four other companies did not provide detailed information about ties with local suppliers.

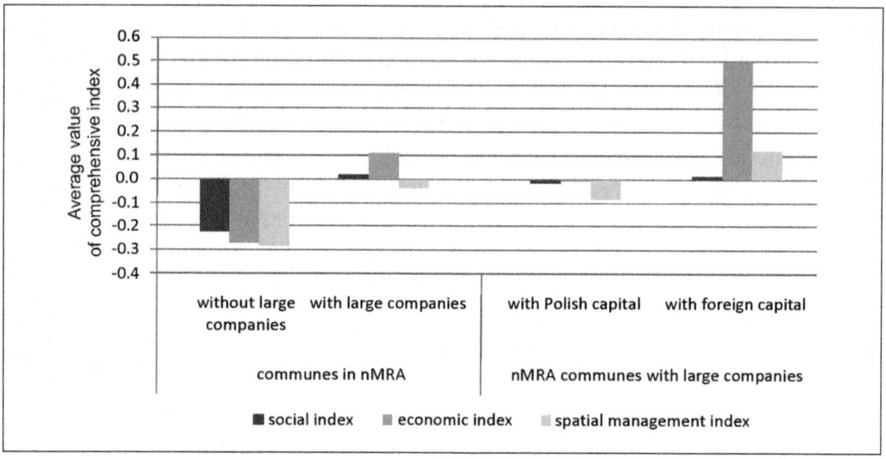

Figure 7.5 Percentage of parts and other materials purchased by the studied companies, based on point of origin

Note: representatives of Depolex and ZL Orzeł considered the areas of Leszno and Jelenia Góra to be the area of the local poviat. In fact they are separate cities with the rights of poviat.
Source: own studies.

is Adros and, in particular, Nicols and Torex, these were the only local services used by such companies. Other business service needs are satisfied primarily by provincial cities. In this respect, the situation at Invado and ZL Orzeł is clearly different. Ties with local service providers are not limited to transportation companies, but also encompass other types of services including knowledge-based services, employee training, and advertising. In the case of both companies, the number of employees of local service providers is much higher than that from non-local companies.

The total number of employees of local companies providing services to large enterprises is the largest in the case of Invado—47 employees; subsequently, Adros—17, ZL Orzeł—11, Torex—10, and Nicols—only 7.

In the last case, the degree of local ties is slightly greater, which is determined by cooperation with workers in cottage industry (area between Trawniki and Lublin) providing services for the company.

From the information provided above, it is possible to draw the conclusion that the impact of large companies on the level of local business activity is quite limited. It is characteristic that the impact is more perceptible in the case of companies with stronger ties to the local community due to the fact that the owners (for example Invado) live in the same area and there is a long tradition of business operation and strong local specialization (ZL Orzeł). The situation is completely different in the case of companies with foreign capital. On the one hand, foreign companies strongly rely on local employees, and on the other, almost completely overlook cooperation with local businesses—suppliers of parts, materials, and services. In the last case, this is most often limited to the use of local transportation companies. Sometimes, weak ties between a large company and its local community are produced by the peripheral nature of an area, which is additionally linked to historically low levels of business activity in a given area. Consequently, few local companies can provide goods and/or services to newly arrived large business enterprises. This is what happened in the town of Małdyty. The town's Kaczkan Company is forced to use non-local service providers, as the local area does not have needed service providers, and local residents do not seem to be able to establish new businesses.

Conclusion

In non-metropolitan rural areas of Poland, there are over 220 large companies whose impact on local communities is variable and its magnitude is determined by several characteristics of enterprises including the key issue of origin of capital. Companies with majority foreign capital constitute 21 percent of all large companies in nMRAs. Foreign-owned companies tend to be located closer to large cities than their Polish counterparts. A characteristic feature of this group is also a higher level of technological advancement in comparison to domestic firms.

A detailed analysis of the level of development of communes in nMRAs depending on the presence of large companies and the origin of capital indicates a clearly higher level of social, economic, and infrastructural development in communes with large foreign companies versus communes with large Polish companies. This difference is even clearer in the case of communes without any large companies or institutions. Research has shown that it is not foreign companies that stimulate local economic growth, but rather foreign companies simply establish plants in areas that are already more developed.

The research has confirmed, to some extent, what has been written in the non-Polish literature about the negligibly beneficial impact of foreign companies on local development. Stronger embeddedness has been observed in the case of many Polish companies. Weak ties with local economic entities were ascertained in the case of companies with foreign capital. In some cases, weak ties are the result of the functional nature of a given geographic area as well as historically low levels of business activity in that given area. In effect, a large business enterprise may not find any suitable service providers in its immediate vicinity.

Similar conclusions have been formulated by Bolesław Domański (2010: 269), who observes that for the strength of regional ties of foreign companies:

> the features of the region itself are of primary importance, and which determine the capacity for seizing multiplier effects, i.e. satisfying business investor demand. Regions with a higher level of development and higher diversification

of the economy may draw greater benefits from multiplier effects due to the presence of foreign investors.

In reference to Perroux's *growth pole theory* (1955), it is necessary to state that only the largest, most resilient, and economically strongest companies may play an important role in the development of rural areas. A condition necessary for such phenomenon is assigning properly significant role to the place of conducting business, manifested by taking care of broadly understood interests of the local community. This stance is only possible in the case of companies that are strongly embedded in the local community, and which have a clear sense of being a "local citizen."

References

Anderson, D., Tyler, P. and McCallion, T. 2005. Developing the rural dimension of business-support policy. *Environment and Planning C: Government and Policy*, 23(4), 519–36.

Bank of Local Data from the Central Statistical Office. 2010. Available at: http://www.stat.gov.pl/bdl/app/strona.html?p_name=indeks [accessed: August 15, 2010].

Burbach, R., Nunez, O. and Kagarlitsky, B. 1997. *Globalization and its Discontents: The Rise of Postmodern Socialism*. London: Pluto.

Dej M., 2009, Large enterprises as a factor triggering social and economical changes in the polish countryside outside the metropolitan areas, *Revija za geografijo—Journal for Geography*, 4–2, 87–102.

Domański, B. 2001. *Kapitał zagraniczny w przemyśle Polski: Prawidłowości rozmieszczenia, uwarunkowania i skutki*. Kraków: Institute of Geography and Spatial Management UJ.

Domański, B. 2004. Local and regional embededdness of foreign industrial investors in Poland. *Prace Geograficzne UJ*, Institute of Geography and Spatial Management UJ, 117, 37–54.

Domański, B. 2010. Kapitał zagraniczny a rozwój polskich regionów, in *Przekształcenia struktur regionalnych—aspekty społeczne, ekonomiczne i przyrodnicze*, edited by S. Ciok and P. Migoń. Wrocław: Institute of Geography and Regional Development, Instytut Geografii i Rozwoju Regionalnego, Uniwersytet Wrocławski, 263–74.

Domański, B. and Gwosdz, K. (ed.) 2005. *Dziesięć lat doświadczeń pierwszej polskiej specjalnej strefy ekonomicznej. Mielec 1995–2005*. Kraków: Instytut Geografii i Gospodarki Przestrzennej UJ, ARP o. Mielec.

Grabher, G. 1994. The disembedded regional economy: the transformation of East German industrial complexes into western enclaves, in *Globalization, Institutions, and Regional Development in Europe*, edited by A. Amin and N. Thrift. Oxford: Oxford University Press, 177–95.

Hardy, J., 1998. Cathedrals in the desert? Transnationals, corporate strategy and locality in Wrocław. *Regional Studies*, 32(7), 639–52.

Keeble, D. and Tyler, P. 1995. Enterprising behaviour and the urban–rural shift. *Urban Studies*, 32(6), 975–97.

Micek, G., Działek, J. and Górecki J. 2011. The discourse and realities of offshore business services to Kraków. *European Planning Studies*, 9, 1651–68.

Nauka i technika w Polsce w 2009 roku. 2011. Warszawa: Główny Urząd Statystyczny.

Oinas, P. 1997. On the socio-spatial embeddedness of business firms, *Erdkunde*, 51.

Oinas, P. 1998, *The Embedded Firm? Prelude for a Revived Geography of Enterprise*. Helsinki: Helsinki School of Economics and Business Administration.

Perroux, F. 1955. Note sur la notion de pôle de croissance. *Economie Appliquée*, VIII, 1–2, 307–20.

Smętkowski, M., Jałowiecki, B. and Gorzelak, G. 2008. *Diagnoza problemów rozwoju obszarów metropolitalnych i rekomendacja delimitacji obszarów metropolitalnych w Polsce*. Warszawa: Centrum Europejskich Studiów Regionalnych i Lokalnych EUROREG, UW.

Smith, D. 1972. Geography and social indicators. *South African Geographical Journal* 54, 43–57.

Stryjakiewicz T. (ed.). 2004. *Wpływ inwestorów zagranicznych na rozwój regionalny i lokalny na przykładzie GlaxoSmithKline Pharmaceuticals S.A. w Poznaniu*. Poznań: Bogucki Wydawnictwo Naukowe.

Chapter 8
Beyond Localness: Development Dimensions of the Niepołomice Investment Zone

Wojciech Jarczewski, Maciej Huculak, Magdalena Dej, Łukasz Sykała

Introduction

Niepołomice is a small town of approx. 12,000 inhabitants located in the vicinity of Cracow (about 10 km from the city centre) within the borders of the Cracow Metropolitan Area (CMA). It has been perceived as an example of local economic success since the beginning of system transformation in Poland (after 1989). The reasons for this success are primarily to be found in the consistent, comprehensive pro-investment policy.

In spite of difficulties in the initial phase of transformation, which took place in the first half of the 1990s, related to a radical decrease in employment in Niepołomice's companies, the town managed to reverse the unfavourable trends; the impulse for change was development of the Niepołomice Investment Zone (NIZ). The effects of 20 years of multi-dimensional pro-investment activities are 60 companies operating in the industrial zone, which invested approximately USD 470 million in the commune. These companies, by employing approximately 4,600 people, contributed to the limitation of unemployment problems not only in the area of the commune, but also in Wieliczka County; what is more, they influenced, in a perceptible manner, Cracow's labour market. The varied structure of companies operating in the zone, which is a result of a conscious policy, led to a situation where, apart from large plants of international concerns, such as Coca-Cola, MAN and Royal Canin, there are also smaller, Polish and local companies.

The rapid economic development of Niepołomice is reflected in the communes' high tax revenues which allowed for the implementation of a series of important communal investments.

The purpose of this chapter is to determine and measure the success of Niepołomice at various spatial scales, especially with respect to the local milieu (understood as the area of the commune) – Cracow Metropolitan Area. Three main research questions were posed:

1. What are the economic and environmental consequences of development of the industrial zone in Niepołomice?

2. How does the development of the Niepołomice Investment Zone impact the inhabitants of Niepołomice and other inhabitants of the CMA?
3. Does it make sense for the inhabitants of Niepołomice to continue expanding activity within the industrial zone considering that they bear the increasingly unfavourable consequences of the functioning of the industrial zone, whereas the benefits 'leak' to the outside?

The chapter presents the brief characteristics of the Niepołomice zone, the economic, spatial and environmental effects of the existence of the NIZ, along with the benefits and drawbacks resulting from the operation of the zone at various spatial scales. These investigations are concluded with a discussion about the sense of further extension of activity in the NIZ from the point of view of Niepołomice inhabitants, as well as from the perspective of the CMA.

Methodological Remarks

In the period between February and August 2010, field studies were conducted in the course of which structured interviews and questionnaires were used. Among companies operating in the NIZ, 34 agreed to participate in the study (structured interviews). In total, these companies employed 3,800 out of the 4,600 people working in NIZ companies. In the course of each meeting, the respondents were asked to complete a questionnaire on basic information about the company. Additionally, to obtain broader knowledge on the processes taking place in the commune, interviews were conducted with representatives of widely-understood institutional facilities: the former and current mayor of Niepołomice, representatives of the police, the county labour office, environmental protection bodies, social assistance, cultural units, schools, the Royal Castle foundation, the forest inspectorate, SEZ Cracow Technology Park and others.

A separate questionnaire was addressed to the commune inhabitants who were asked to express their opinions about the zone development – over 460 people took part in the study. Additionally, data from financial statements about the volume of income tax paid were used for analysis of the economic impact of companies in the Niepołomice Investment Zone. On this basis, the volume of tax revenues (corporate income tax – CIT) was estimated from companies operating in the zone which goes to the budget of the Niepołomice Commune, Wieliczka County, as well as Małopolska Province. Data made available by the County Labour Office in Wieliczka was also used for analysis of the market situation.

Impact and Embeddedness of Companies – Recent Studies

Discussions on the local economic success as an effect of the pro-investment policy and, in consequence, operation of economic entities, are based in the context of previous studies on the impact of companies on the local environment.

Literature devoted to the impact of economic entities on the local environment is relatively poor and such studies are frequently combined with issues regarding embeddedness, which is a broader term in relation to companies' impact. Impact takes into account unilateral actions, that is the activities of companies focused on the environment; on the other hand, embeddedness refers to bilateral actions. There are numerous ways of depicting the embeddedness of economic activity (inter alia Oinas 1997, 1998), yet all of them take into account mutual relations between the analysed entity and the environment (natural, social and economic environment).

A very important conclusion regarding the relations of foreign companies with local economic entities is the statement of Bolesław Domański (2010: 269) who notices that with respect to the power of regional ties of foreign companies:

> the features of the region itself are of significant weight; they determine the capacity for capturing multiplier effects, i.e. satisfying the investors' demands. Regions with higher level of development and greater diversification of economy may draw greater benefits from multiplier effects of foreign investors' presence. Relations of local companies with investors from outside are often determined by the functional nature of the place of location and such features as the level of local entrepreneurship.

At the same time, the impact of companies, as Stryjakiewicz (2004) noted using the example of foreign investors, results from the operation strategy which is adopted; Stryjakiewicz differentiates three types of spatial expansion of multinational companies from the point of view of their impact on the regional and local economy: a type oriented towards the use of local resources, a type of 'export platforms' where location abroad is aimed at capturing local outlet markets and a type connecting global corporate strategies with 'embeddedness' in the local economy.

There are numerous studies on the embeddedness of foreign economic entities in reference to countries in Central and Eastern Europe, including Poland. It is worth noting that in studies by foreign researchers, it is possible to perceive fewer positive aspects of such functioning and slight embeddedness of companies (Grabher 1994; Hardy 1998, Burbach, Nunez and Kagarlitsky 1997) than in the case of Polish researchers. Domański (2004) notes that the operations of the majority of large producers from abroad have a positive impact on the competitiveness of a location, mainly by improving the prospects of local companies and increasing qualifications. Such entities may be perceived as local citizens as a consequence of paying taxes, compliance with environmental provisions and good labour relations (Domański 2004: 16).

It is worth emphasizing the fact that the success of Niepołomice is not a consequence of the impact of only one company; it is an effect of development of the entire investment zone. This development is directly related to its location in the vicinity of Cracow and the possibility of using its resources. Ocipka (1999), in reference to investment zones, lists the following elements which determine their success: transportation accessibility, energy resources, good technical infrastructure, level of education of human resources and access to social infrastructure. Thus, zones located near or directly within large municipal and industrial agglomerations will have the greatest development chances (Domański and Gwosdz 2005). Taking the above statements into account, it is necessary to think about the inter-dependence of Niepołomice and Cracow with respect to the economic success of Niepołomice which, thanks to the vicinity of the metropolis, has conditions conducive to development of the investment zone and Cracow, which draws a lot of benefits from the presence of large industrial plants within its metropolitan area.

Characteristics of the Niepołomice Zone

Niepołomice is a small municipal/ rural commune (25,000 inhabitants, including 12,000 in the town of Niepołomice), adjoining Cracow (mainly the Nowa Huta district) via the Vistula River from the south-eastern direction (Figure 8.1). In the 20 years of development of self-government in Poland (since 1990), Niepołomice has become the national symbol of professionalism and efficiency in attracting investors. The difficulties of the initial phase of social economic transformation of the country which took place in the first half of the 1990s were also experienced in this small Cracovian commune. Employment in all large- and medium-sized companies operating in Niepołomice was radically reduced. The situation was aggravated by an increase in unemployment resulting from dismissals of dual-profession inhabitants of the commune from the Tadeusz Sendzimir Steelworks located in the direct vicinity of Niepołomice.

At the same time, as a result of local government elections in 1990, Stanisław Kracik was chosen as the Niepołomice mayor. Along with a group of co-workers, he commenced a process of social and economic modernization of the commune and one of the foundations of the changes was the development of the Niepołomice Investment Zone (NIZ) – an area where new investments were located. The effects of 20 years of multi-dimensional pro-investment activities are 60 companies operating in the industrial zone, which invested approximately USD 470 million in the commune (Jarczewski and Huculak 2011). These companies, by employing almost 4,600 people, contributed to a decrease of problems related to unemployment, not only in the area of the commune, but also in Wieliczka County, thereby noticeably impacting the labour market of the city of Cracow. The diversified structure of companies functioning in the area of the zone, resulting from a conscious policy, led to a situation where, apart from large plants of international concerns, such as Coca-Cola, MAN and Royal Canin, there are also smaller domestic and local companies operating there.

Rapid economic development of Niepołomice was reflected in high revenues for the commune, which allowed for the implementation of a number of important municipal investments (such as the development and modernization of the local road system, development of a sewage treatment plant and construction of a sewage system). In October 2009, Stanisław Kracik resigned from being the mayor after 19 years and was appointed to the position of the Małopolska Province Governor.

Development of the Niepołomice Investment Zone (NIZ) may be divided into four basic stages:

- Stage I (1990–1994). The period of the downfall of state enterprises and development of individual, small private production companies; a time of quick growth of unemployment in the commune. This period ended with opening of the first large and very prestigious investment – the production plant of Coca-Cola.
- Stage II (1995–1999) the period of influx of new investors to the commune who located themselves almost exclusively in post-industrial areas (brownfield), which constituted the majority of the commune's resources. In this period, mainly production companies were located in Niepołomice, even though this stage ended when a golf club started to operate there in 1999.
- Stage III (2000–2005). This is the time when many new, large industrial investments were started in the area of the commune, including the emergence of the currently largest employer in the commune, Oknoplast Kraków. In parallel to these processes, intensification of activity conducted by companies present in the investment zone since the 1990s was perceptible.
- Stage IV, after 2005. Intensified influx of investors to the new investment areas (greenfield). Since 2001, the commune had been preparing itself to accept a new large investment. These efforts were completed with a decision of MAN Trucks to build an assembly plant for trucks in Niepołomice in 2005. Along with this investment, pro-investment efforts in the commune started to be supported by the SEZ Cracow Technology Park and the Małopolska Industrial Parks – a local government company. In this period, more and more service companies started to locate their headquarters in the commune and develop their operation.

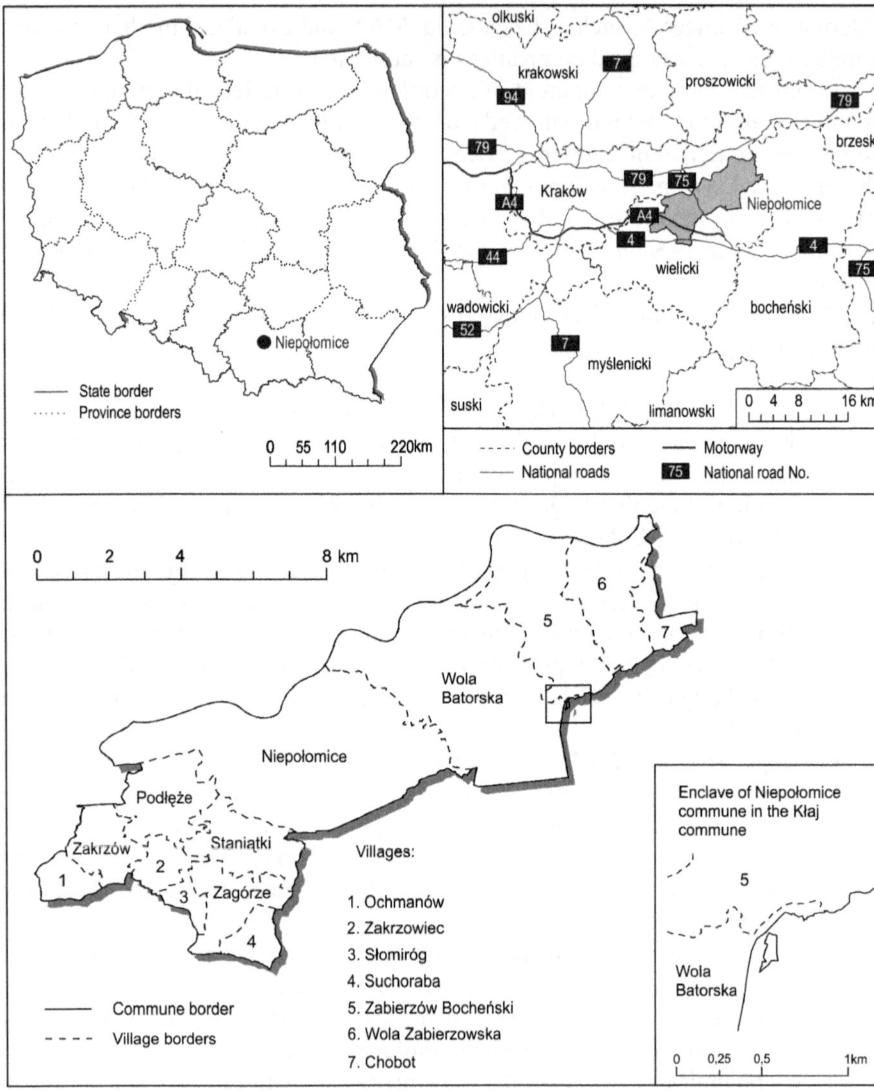

Figure 8.1 Location of Niepołomice within the borders of Poland, the Małopolska Province, direct resources of Cracow and administrative division of the commune

Source: own study.

Effects of the New Investment Zone in Niepołomice at the local and CMA Scale

Labour Market

The establishment and development of the Niepołomice Investment Zone contributed to definite improvement for the local labour market – in 'zone' companies, approximately 4,600 jobs were created. As a result, unemployment in the commune, which was very high in the first half of the 1990s, has clearly decreased and is currently among the lowest in Małopolska Province. Companies that commence their operation in Niepołomice offer wages which do not clearly differ from the average value for the entire region.

Many employees of Niepołomice companies reside outside the commune, which considering the relatively small local labour market is the only possible way the new companies can constant develop. The area of the commune is inhabited by slightly over 24 per cent of employees of companies operating in the Niepołomice Investment Zone. Most jobs are occupied by the inhabitants of Cracow – over 40 per cent. In the light of the studies conducted, the majority of people commute to work in the plants of the Niepołomice Investment Zone within a distance of 30 km from the town. A dominant role in this respect is played by Cracow which, along with the Niepołomice commune, is responsible for 2/3 of commuting employees (Figure 8.2).

The nature of the operations pursued by the Niepołomice companies results in the fact that the labour market does not depend on one employer, which reduces its susceptibility to crisis-type phenomena. Employees from liquidated companies or companies that reduce employment may find new work relatively easily in other developing companies.

In order to better describe the issues related to the impact of the plants operating in the Niepołomice Investment Zone on the structure of supply of new jobs, data collected by the County Labour Office in Wieliczka was used. The data made available to the authors, which derive from the IT system of the County Labour Office in Wieliczka, confirm the fact which was emphasized in talks conducted with entrepreneurs from the NIZ area. The observed phenomenon is that not all companies use the services of the County Labour Office in Wieliczka, and some use such services in exceptional situations, for example, in the case of significant personnel shortages in the labour market. Thus, in the breakdown which is the basis of this study, the largest investor in the NIZ (for example the German company MAN Trucks) is missing; the presence of the largest employer (for instance Oknoplast Kraków) is only symbolic, along with other, medium-sized employers (for example DHL, Royal Canin and others). In spite of the above-mentioned deficits, the database of the County Labour Office in Wieliczka is the best depiction of the situation in the labour market in Wieliczka County with respect to changes in demand for individual professions. Out of almost 14,000 records in the database regarding work offers for specific positions submitted by companies to the County

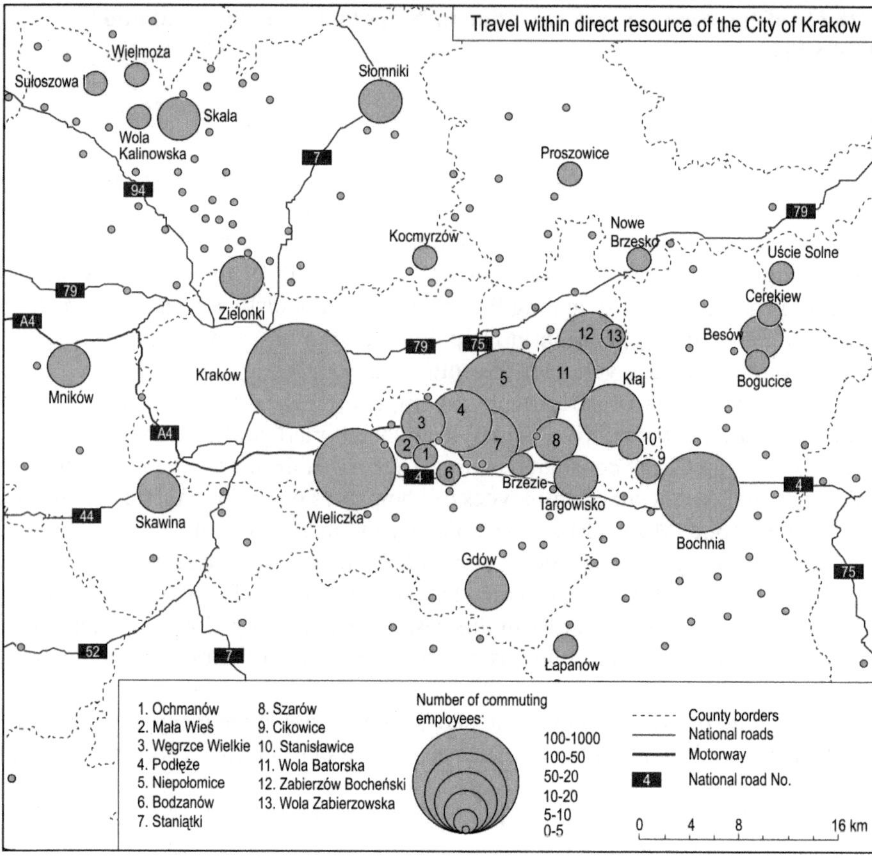

Figure 8.2 People commuting to work in the Niepołomice Investment Zone in 2010

Source: own study on the basis of data made available by the companies in the NIZ.

Labour Office in Wieliczka between 1991 and 2010,[1] a group of 31 companies was separated, which operated at that time in the Niepołomice Investment Zone. The number of positions offered by companies from the NIZ ranged, in individual years, from 29 to 466. The total number of jobs offered amounted to 2,661 in 115 professions, out of which, in the area of Wieliczka County, 1,400 employees who were also inhabitants of the county were accepted for work within the scope of work agency services offered by the County Labour Office in Wieliczka. In the same period, in Wieliczka County, employers were searching for approximately 60,000 employees in 757 various professions, employing approximately 30,000

1 The year 2010 represents the status as of October 14.

county inhabitants. In the entire county, such professions as sellers (8.5 per cent offers), other office service employees (6.9 per cent), farm workers (5.7 per cent), auxiliary workers in the processing industry (5.0 per cent) and construction workers (3.6 per cent) were dominant. The jobs offered by Niepołomice companies, on account of their specialization within the scope of production activity, obviously significantly differed from data for the entire county. Half of the employees whom employers were seeking were in professions such as: auxiliary workers in the processing industry (5.0 per cent), construction workers (18.0 per cent), persons without profession (11.2 per cent), farm workers (7.3 per cent), warehouse keepers (5.5per cent), truck drivers (4.7 per cent) and other office service employees (3.2 per cent). From the point of view of the education preferred by employers, the dominance of jobs for persons with primary and vocational education is clearly visible. However, in the entire county the share of jobs offered to people with this type of education systematically decreased between 1994 and 2007 from approximately 90 per cent to approximately 60 per cent. The trend of the analysed phenomenon for the companies from the NIZ is much more rapid.

Detailed analysis of jobs of the County Labour Office in Wieliczka allows for diversification of development stages of the Niepołomice Investment Zone from the labour market point of view (Figure 8.3). The number of employees sought by the companies from the NIZ, via the County Labour Office in Wieliczka, constitutes, according to the authors, a good exemplification of the historical development processes of the Niepołomice zone:

1. The first development stage of the industrial zone in the Niepołomice commune was the economic boom after the rapid drop in industrial production in the country. This was the period between 1994 and 1997 and, in the case of Niepołomice, it also included the year 1995 when the number of employees sought did not differ from the lowest values for the entire period. The majority of offers and, in consequence, employees accepted for work were in two of the 'old' plants in the industrial zone, that is Garbarnia ADA and Niepołomickie Zakłady Drobiarskie, as well as new, large foreign investors, such as Coca-Cola, Dresdner Fernsterbau along with its co-operator, Glaspol and Kamm-Bauelemente. An additional factor was the transfer, in 1996, of cured meat production from Igołomia by Stastnik Polska. In total, out of 504 employees sought, 320 people got work in Niepołomice plants in the course of 4 years.
2. The slow-down period in the labour market was between 1998 and 2001. In the course of the subsequent four years, employers accepted 89 out of 161 people. The majority (55 per cent) of people employed at that time were working in the largest plants in the zone, that is Coca-Cola, NZD, Dresdner Fernsterbau, Marseille, Pharma-C-Food, Stastnik and a new investor Staco Polska, who in 1999 transferred its production plant to Niepołomice from Minków (Liszki commune). In this period, it was possible to see the first symptoms of an acceleration in the development of existing companies, for example Marseille and Pharma-C-Food.

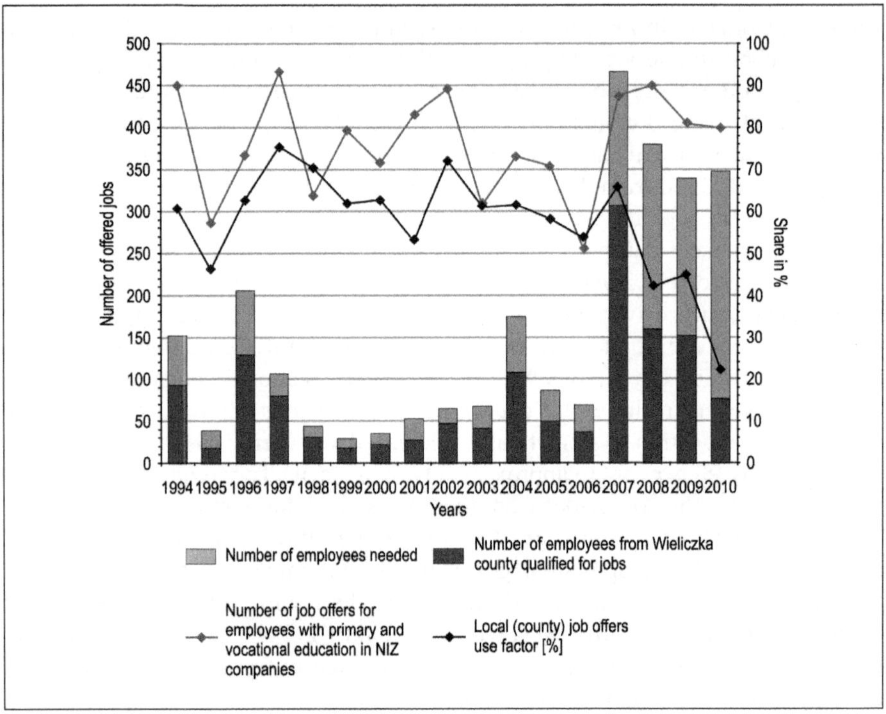

Figure 8.3 Data regarding the activity of the County Labour Office in Wieliczka between 1994 and 2010

Source: own elaboration based on the basis of data obtained from County Labour Office in Wieliczka.

3. The situation improved between 2002 and 2006. The peak of demand for employees took place in 2004, when the employee need declared by the Niepołomice companies was close to the record levels of 1994 (153) and 1996 (206), reaching 175 employees. In this year, high demand for new employees was related to the expansion of existing companies, such as Coca-Cola, Pharma C-Food, DFB, Cedrob, Stastnik, Staco, Marseille, or Oknoplast Kraków, which has been present in the zone for a shorter period of time, from 2000. This period was also marked by development of non-production functions, including administrative functions, visible in the plants of Stastnik Polska, Coca-Cola and Oknoplast; as well as an increase in demand for specialist workers with technical education, including higher education. The last group includes both machine operators and production engineers. There are also jobs on offer for middle and higher grades of

management. In total, in this period 19 people with higher education from Wieliczka County started to work in the zone.[2]

4. The boom period in the labour market was in 2007 and 2008. The group of companies operating in the area of the industrial zone which voiced their demand for at least several dozen employees, included Coca-Cola, Cedrob, Staco, Stastnik, Oknoplast-Kraków, Nycz Intertrade, and Amcor White Cap and they were joined by new plants such as Polynt, Madro, n-Vision and Terra Polska. In the course of this short period of only two years, selected Niepołomice companies voiced their demand for 846 employees, employing 468 people from the county area. This period drained the resources of the local labour market. The index of local use of labour offers which, excluding the year 1995, ranged between 50 and 75 per cent, dropping to the level of approximately 20 per cent in 2010.
5. The years 2009 and 2010 are the slow-down period in the labour market of Niepołomice. They are characterised by a clear drop in the number of new jobs in comparison to tendencies observed in the entire county. In this period, the lower drop in the number of jobs offered was mainly influenced by the development of three companies: Cedrob, Stastnik and Marseille. At the same time, it was possible to notice a further drain on local (county) labour market resources. The index of local use of jobs offered reached its historical minimum with a value of 22 per cent.

Tax Receipts

The best exponent of the commune's affluence is the volume of tax receipts (such as income taxes, property tax). In comparison to other urban and rural communes in Poland, where the number of inhabitants, similarly to Niepołomice, does not exceed 30,000, the tax revenues of Niepołomice are strong. With respect to the volume of tax receipts in 2010 in reference to all basic types of taxes: real estate, transportation, share in PIT (personal income tax) and CIT (corporate income tax), Niepołomice was in the lead (Table 8.1). Particularly favourable was the situation with respect to the real estate tax – its value in 2010 placed Niepołomice in 9th position among 558 urban and rural communes of this size (EUR 4.2 million with the average value for the entire group on the level of EUR 1.1 million). What is more, the share in corporate income tax (CIT), a decisive degree of which was shaped by companies operating in the NIZ, locates Niepołomice in 11th position in Poland (approximately EUR 0.5 million in relation to the average of EUR 0.07 million for the compared group of entities). In this respect, the affluence of inhabitants measured by the receipts of the commune budget from personal income tax (PIT) is also positive. This value, at the level of EUR 3.0 million, places Niepołomice in the 26th position among all smallest urban and rural communes in the country. Additionally, it is worth paying attention to the fact that very high values of taxes are coupled with their above-average dynamics observed in the last decade.

2 Between 1994 and 2001, the number was 10 people.

Table 8.1 Tax receipts of the Niepołomice commune in comparison to urban and rural communes of Poland below 30,000 inhabitants

Unit / measure	Type of tax							
	Real estate		Transportation		Share in PIT		Share in CIT	
	Value, 2010 (in EUR,000)	Dynamics 2000–2010 (%)	Value, 2010 (in EUR,000)	Dynamics 2000–2010 (%)	Value, 2010 (in EUR,000)	Dynamics 2000–2010 (%)	Value, 2010 (in EUR,000)	Dynamics 2000–2010 (%)
Niepołomice	4160.9	400.8	113.3	235.8	2951.7	287.7	475.8	291.3
Average for urban and rural communes below 30,000 inhabitants	1121. 6	248.8	64.4	246.3	1081.2	220.2	71.0	487.9
Position of Niepołomice among all urban and rural communes below 30,000 inhabitants	9/558	32/533	79/558	213/533	26/558	48/533	11/558	193/533

Source: own calculations based on the Bank of Local Data (2012).

Using the financial statements of all companies operating in the investment zone of Niepołomice, values of taxes (CIT) paid by these companies were analysed and subsequently summary values were estimated which go to the budget of Niepołomice Commune, Wieliczka County and Małopolska Province.[3]

According to such estimates, between 2006 and 2010, the budget of Niepołomice Commune received approximately EUR 2 million, out of which almost half was in one year (2008), which is a consequence of a very high profit made by MAN (Figure 8.4). In the same period, taxes from NIZ companies supplied the budget of Małopolska Province with approximately EUR 4.1 million.

The significance of companies from the Niepołomice Zone for local development is clearly evidenced by the share of NIZ companies in the total volume of tax receipts on account of share in CIT at various spatial scales (Figure 8.5). In the period analysed between 2006 and 2010, such receipts constituted more than a half of total receipts of this kind at the scale of the commune (in 2008, as much as 80 per cent), whereas at the scale of the county, this share was also significant (from 22.4 per cent in 2006 to 56.3 per cent in 2009).

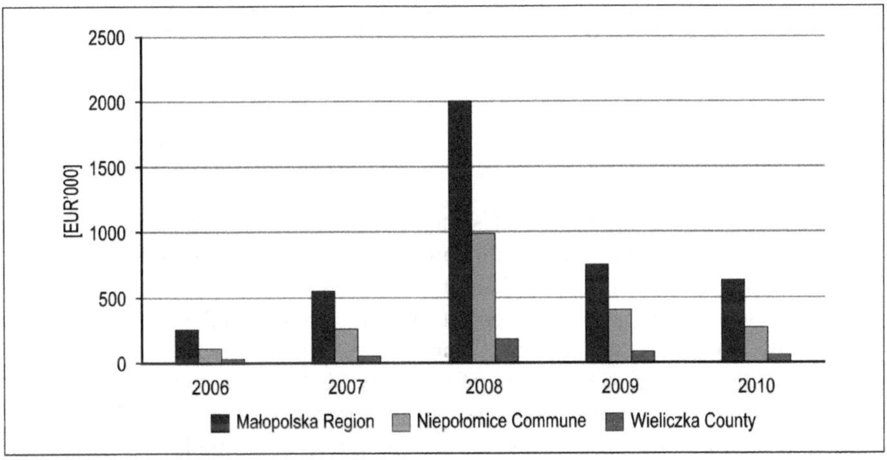

Figure 8.4 Volume of tax receipts to the budget of the Province, the Commune and the County on account of share in CIT from companies operating in the area of the Niepołomice Investment Zone between 2006 and 2010

Source: own study on the basis of financial statements of companies.

3 The estimation was conducted on the basis of provisions of the Act on Income of Self-Government Units of November 13, 2003 (Journal of Laws 2003, No. 203, Item 1966), which stipulates that in the case of corporate income tax (CIT), the share of the commune where the economic entity is registered amounts to 6.71 per cent, the county 1.4 per cent, and the province 14.75 per cent.

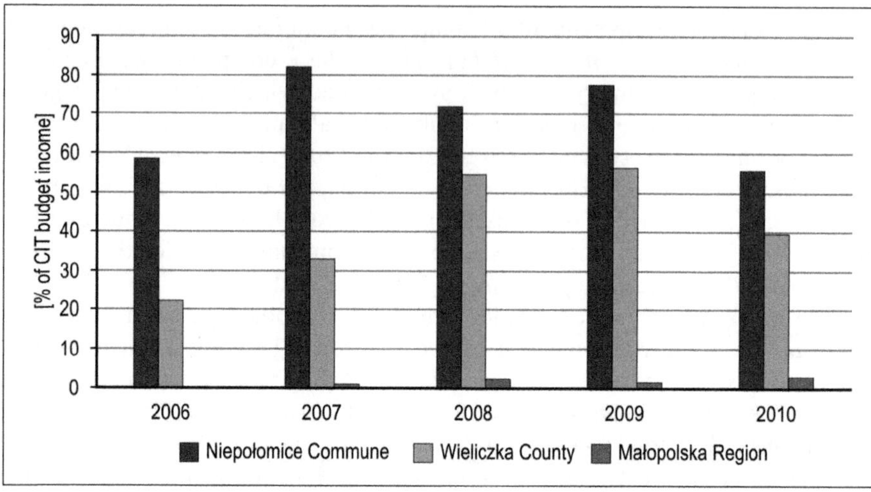

Figure 8.5 Share of companies in the Niepołomice Investment Zone in the total volume of tax receipts from share in CIT in the budget of the commune, the county and the province between 2006 and 2010

Source: own study on the basis of financial statements of companies.

Image of the Commune

Niepołomice entered the period of system transformation in Poland in 1990 as a typical agricultural and industrial commune located in the vicinity of Cracow. Direct vicinity of Nowa Huta, a district of Cracow, where the Lenin Steelworks had a very negative impact on the natural environment, resulted in the fact that very few people, excluding the inhabitants, considered Niepołomice an interesting place from a residential or business point of view. By 2010, the image of Niepołomice has been subjected to radical changes, in Małopolska and in broader terms. In a significant part of southern Poland, the commune developed a very attractive image, related mainly to the pro-investment policy.

The image of Niepołomice as a commune attractive for investors is a result of numerous activities implemented in the first twenty years of self-governing. The local authorities managed to create the brand of Niepołomice as a place friendly to investors in direct vicinity of Cracow in line with the concept: *Think about Cracow, invest in Niepołomice.* The image was created as a result of professional and consistent activities related to the preparation of offers and servicing of investors. The positive message was intensified by other activities and events related to commune development. In this context, it is worth emphasizing the special significance of Mayor Stanisław Kracik for development of the commune's image. By holding the office for 19 years (between 1990 and 2009), he became one of the most recognizable mayors in the country. By developing the commune in a non-standard and innovative manner, he worked out an image of a specialist in attracting investors.

Among the most important promotional tools which built the pro-investment image of Niepołomice, it is necessary to list indirect promotion. The message of Niepołomice was supported by various types of partners, who, within the scope of their activities, provided information about the attractive offers of the commune. The most important institutions supporting the promotion of Niepołomice are: the Polish Information and Foreign Investment Agency, Małopolska Regional Investor Servicing Centre, SEZ Cracow Technology Park and Małopolska Industrial Parks.

A special role in development of investment zones in Niepołomice was played by investors who already operated there. The process of procuring new companies was greatly accelerated by appearance of the new Coca-Cola production plant in Niepołomice. The location of this most recognizable brand in the world (*The World's* ... 2006) diametrically changed the contemporary image of the commune, providing it with *investatur* – an opinion that it was worth to invest there. The positive image was supported, on a much greater scale, by the location in the commune of the MAN Trucks assembly plant in 2005.

Stances of Niepołomice Inhabitants in regards to the Establishment and Development of the NIZ

For the purpose of getting to know the opinion of the inhabitants of the commune and the town of Niepołomice with respect to economic development and perception of industrial activity conducted in the area of the commune, a questionnaire regarding stances with respect to the natural environment was circulated. The studies were conducted on the basis of a questionnaire entitled *The Niepołomice Commune in the Eyes of Its Inhabitants.* It was addressed to people older than 18 and distributed among the inhabitants thanks to cooperation with primary schools in the examined locations. In total, 460 completed questionnaires were received. The manner of questionnaire distribution had a significant impact on the structure of respondents. A definite majority were middle age people (31–50 years of age) (75.9 per cent), whose children attend schools. At the same time, this group is characterized by greatest professional activity, which is significant from the point of view of this study. The manner of conducting the study also influenced the share of men and women in the total number of respondents. Women constituted over 70 per cent of respondents. This fact has to be taken into account when analysing the results if it turned out that sex was a variable that has significant impact on answers. Over 2/3 of respondents had at least secondary education. The most numerous group were people with secondary education (38 per cent). About 24 per cent of respondents declared having higher education. The group of people with basic vocational education was also numerous (27 per cent). Among people who specified time of residence in the Niepołomice commune, the largest group (45 per cent) were people who lived there for 26 years or more. About 25 per cent of people declared the time of residence not longer than 10 years, and further 30 per cent between 11 and 30 years.

A definite majority of respondents considered the existence of the zone an important factor of the commune's success (almost 90 per cent of respondents). What is more, a significant majority of respondents believe that the zone is very important from the point of view of the labour market, whereas almost half definitely believes so. Over 80 per cent of respondents agreed with a statement that the Niepołomice Industrial Zone is a showcase of the commune. The respondents show slightly less enthusiasm with respect to the idea of further extension of the zone, even though a definite 'no' was voiced solely by every third respondent, and decisive opponents of the zone extension constituted 13 per cent. In reference to negative statements regarding the zone (it is harmful for the environment, it limits development of other activities and services), opinions were more divided. The percentage of people who believe that the zone is harmful for the environment reached 44 per cent. The majority of respondents do not agree with the negative impact of the zone on development of other activities and services in the commune, even though a definite opinion in this respect was only voiced by every fifth respondent. The greatest diversification of answers referred to the statement that the Niepołomice Industrial Zone differentiates development of individual parts of the commune. In the case of this question, a high percentage of answers in the category 'I cannot say' was noted, which may suggest that expressing an opinion on a statement formulated in this manner posed certain difficulties. This is how the significant variety of answers may also be explained.

The inhabitants of Niepołomice have positive stances with respect to industrialization as such, as well as in reference to industrialization of their own commune and town. Industrial activity in the commune is evaluated best among all examined types of activity. The inhabitants perceive development of industrial activity and location of subsequent investments not only in the categories of improvement of the local labour market, but also as an important factor leading to economic development of the entire commune or an important element building the brand of the commune, creating its positive image. In this sense, new industrial investments constitute an important element of local identity. The inhabitants of the town and the commune of Niepołomice willingly identify themselves with the place perceived from the outside as an example of economic success, and, at the same time, a place with high standard of living.

Concerns of a major part of inhabitants with respect to the negative impact of industrial activity conducted in the commune on the natural environment mainly referred to indirect effects such as, for example, generation of excessive vehicle traffic. Industry as such is not perceived as harmful for the environment (for instance via emission of pollution). In the case of the inhabitants of the Niepołomice commune, it is difficult to talk about any conflict of values along the line: industrial activity – environment. The pro-investment stance of a major part of the respondents does not exclude adoption of a pro-environmental stance at the same time; the high evaluation of industrial activity in the commune is not, in any manner, juxtaposed with a high assessment of natural values and acceptance for increasing the surface areas of protected sites.

However, paradoxically, positive evaluation of industrial activity in the commune results in the lack of support for an increase of the commune's outlays on this branch of economy and resistance to the idea of extension of the existing industrial zones. The inhabitants are quite unanimous as far as opinions expressed in reference to the entire commune are concerned, even though in the case of industrial activity, it is possible to notice certain differentiation depending on location with respect to the concentration areas of industrial activity and work places (within the commune or outside).

Impact of Development of the NIZ on the Natural Environment

Development of the Niepołomice Investment Zone, on account of the scale of the operations there, is related to certain negative impact on the natural environment in the commune. However, it has to be noted that in relation to the number and the size of new investments, this pressure is not particularly visible. It is also necessary to remember that in Niepołomice, development of new companies was accompanied by the closure of older companies which were established in the period of the People's Republic of Poland and which had negative impact on the environment, in particular the tannery and Krakowskie Zakłady Drobiowe (Cracow Poultry Farm).

The profile of companies commencing operation in Niepołomice is mainly industry (with little troublesome impact for the environment) and services, whose impact is most frequently limited to investment plots. The most serious negative impact of development of new companies is related to the building up of green areas and decrease in the size of biologically active areas. As a result, some animals (mainly birds), lose their feeding grounds; problems also emerge with respect to high levels and rapid runoff of surface waters, in particular during heavy rains or thaw. Currently, new investments are implemented on the condition of securing storage reservoirs for water deriving from them. Farsightedness and efficiency of activities, developed jointly by the County Inspectorate for Environmental Protection and the Niepołomice Town and Commune Office with respect to optimizing hydrological conditions within the NIZ, confirmed the absence of any local flooding resulting from obstruction or insufficient flow capacity of the rainwater discharge system during intensive rainfall in May and June 2010.

New enterprises, particularly those that operate in the food industry (Coca-Cola bottling plant, poultry farm), use local water supply sources. The scale of such needs is still slight in comparison to the efficiency and does not threaten exhaustion of resources. However, the fact that the quality of water is worsening, both from Quaternary intake points, as well as Tertiary ones, is worrying. This tendency, even though independent from the entities operating in the area of the NIZ, may influence their operation, inter alia by an increase in the costs of water treatment. This may limit, in a certain respect, the further influx of investors from the food industry to the NIZ.

Within the area of the NIZ, there are some entities involved in waste processing, including hazardous waste. This fact increases a potential threat of air pollution. However, the majority of pollution observed in the commune derives from the Arcelor Mittal steelworks in Cracow, as well as other plants located outside the area of the commune and from local low emissions.

Benefits and Drawbacks Resulting from the Existence of the Zone in Various Dimensions

The Niepołomice Investment Zone was established at the beginning of the 1990s for the purpose of activating the local economy, painfully experiencing the initial period of transformation, and reducing the rapidly growing unemployment. These purposes, inter alia thanks to skilful use of favourable conditions by the local authorities – were, in principle, implemented at the turn of the century. However, after 2000, development of the NIZ was not halted – it even significantly accelerated. In this period, the NIZ more and more clearly started to take over the functions of the investment zone of the Cracow Metropolitan Area.[4]

In 2005, the 'metropolization' process of the NIZ obtained formal and financial confirmation on the part of self-government authorities. For the purpose of attracting (successfully, as it turned out) the German producer of trucks, MAN Trucks, the Małopolska Marshall Office established a special company – Małopolska Industrial Parks. In principle, the only task of this company was to purchase plots of land from private owners, consolidate them and sell them to investors in Niepołomice. At the same time, some investment plots in the commune were encompassed by tax discounts within the mechanism of Special Economic Zones (SEZs).[5]

4 Cracow Metropolitan Area (CMA) does not have any legal personal or a formal basis. So far, its borders have not been clearly defined. In various documents, it covers an area ranging from several dozen Małopolska communes (Małopolska Regional Operating Programme 2007–2013) to over 50 (Management Plan for the Małopolska Province of 2003). However, disregarding the adopted borders, this area complies with basic criteria characteristic for a metropolitan area – as an urban system of multiple settlement units and areas with high degree of urbanization, whose main feature is occurrence of metropolitan functions, as well as functional and economic ties (Markowski and Marszał 2006). The number of inhabitants of this region, depending on the adopted delimitation, oscillates around 1 million.

5 Special Economic Zones in Polish conditions are financial mechanisms which allow for exempting entrepreneurs from CIT in investment plots encompassed by the SEZ.

Currently, the NIZ undoubtedly functions as the most important and largest investment zone in the Cracow Metropolitan Area. This is a location for companies (mainly production, but also service companies) interested in human potential, sub-suppliers and service providers, as well as the outlet market of the CMA. In this respect, questions emerge about benefits, difficulties and problems related to development of the NIZ at various spatial scales which are synthetically presented in Table 8.2.

Table 8.2 Benefits of the Niepołomice Investment Zone in various dimensions (1)

Dimension	**Benefits**	
	Niepołomice	**CMA**
Development of entrepreneurship	Development of NIZ transformed the character of Niepołomice after 1990 from an agricultural/ industrial commune to an industrial and service commune.	New companies located in the NIZ, including large multi-national concerns became an element of a broader process of transformations and modernization of the CMA economy/ model investment offer attracts companies relocating from Cracow, preventing their drainage outside the CMA.
Labour market	Radical decrease in the level of unemployment in the commune.	NIZ is an important labour market for the CMA inhabitants, in particular Cracow and Wieliczka.
Image	Along with development of the NIZ, Niepołomice obtained a very positive image of a dynamic commune, attractive with respect to investments which facilitates procurement of subsequent companies.	Location of new companies in Niepołomice strengthens the image of the CMA as a place that is attractive with respect to investments/ some companies advertise their products under the brand of Cracow, e.g. windows from Cracow.
Population problems	Niepołomice are most rapidly developing town in Małopolska with respect to demographic aspects.	Demographic development of Niepołomice strengthens the human potential of the CMA (quality and quantity) which is conducive to development of a strong metropolis.
Natural environment	Closing down of most burdensome plants – tannery and poultry farm.	–

Source: own study.

Table 8.3 Benefits of the Niepołomice Investment Zone in various dimensions (2)

Dimension	Benefits	
	Niepołomice	CMA
Receipts from taxes (PIT, CIT, real estate)	Radical increase of the commune's affluence as a result of revenues on account of such taxes.	Cracow, Wieliczka, as well as other communes of the CMA obtain additional receipts from PIT, and, additionally, the Wieliczka County and the Małopolska Province have a share in PIT and CIT
Multiplier effects	–	Relatively large closure of multiplier effects – income and supply within the scope of services, thanks to the presence of Cracow and specialist services located in it.
Social sphere	Perceptible acceptance for industrialization processes in the commune.	–
Transport	Modernization of road system – western ring road of the city.	Transfer, along with plants, of a part of truck traffic outside of Cracow.
Other technical infrastructure	Possibility of indirect subsidies for the volume of fees for water and sewage.	–

Source: own study.

According to the breakdown (Tables 8.2 and 8.3), from the perspective of the CMA, development of the NIZ is, in principle, clearly beneficial. Negative phenomena, such as pressure on the natural environment related to urbanization and industrialization of the commune seems to be slight (Table 8.4 and 8.5). Apart from that, in principle the CMA only draws benefits from the functioning of the NIZ.

However, the situation is a bit different in the Niepołomice commune. Undoubtedly, the commune drew significant benefits from development of the NIZ and, as has been indicated – its development is perceived quite positively by the inhabitants. However, the question about the growth limits is being asked more and more often. The labour market which was created in the commune significantly exceeds the local potential. In total, in the NIZ companies, only a quarter of employees are commune inhabitants and in new enterprises, this percentage is even lower. It is also not possible to avoid growing pressure on the natural environment, and it is necessary to remember that the largest forest complex of the CMA is located in Niepołomice, that is the Niepołomicka Forest and its protection is an important objective of the commune. What is more, growing vehicle traffic, in spite of new numerous infrastructural investments, does not make the inhabitants' lives easier.

Table 8.4 Difficulties and problems of the Niepołomice Investment Zone in various dimensions (1)

Dimension	Difficulties and problems	
	Niepołomice	**CMA**
Development of entrepreneurship	In spite of significant enthusiasm of the local government, cooperation among companies in the NIZ is developing quite slowly – some of them function as 'cathedrals in the desert' and unwillingly establish local relations and ties.	Niepołomice, even though in the scale of the CMA it conducted the most efficient pro-investment policy, yet on account of various types of restrictions – is not able to create certain offers for demanding customers and other CMA communes are not always able to prepare attractive offers/ lack of planning of pro-investment policy on the metropolitan scale.
Labour market	Exhaustion of the Niepołomice labour market potential / mismatching of employee qualifications for the needs of new employers.	–
Image	Recognition of Niepołomice is fundamentally limited to southern Poland and in expert circles to the entire Poland. Outside the borders of Poland, investment in the NIZ are identified more with Cracow or with the CMA.	–
Population problems	Increase in the number of inhabitants means an additional increase in pressure on the natural environment and in particular the Niepołomicka Forest.	–

Source: own study.

Taxes received by the commune budget in relation to the functioning of enterprises in the NIZ (the commune's share in CIT, the commune's share in PIT paid by persons registered in the commune, real estate tax, transport tax, and so on) result in the fact that Niepołomice belongs to the most affluent communes in Małopolska.[6] Numerous municipal investments result in the fact that the quality of life of city inhabitants is continuously growing. What is more – the receipts are stable and constantly increasing. On account of the very strong diversification of companies

6 In *Ranking gmin małopolskich* MISTIA 2012, Niepołomice had the 2nd place, where the average annual value of own income per inhabitant PLN 2,680 was the best result in Małopolska.

Table 8.5 Difficulties and problems of the Niepołomice Investment Zone in various dimensions (2)

Dimension	Difficulties and problems	
	Niepołomice	**CMA**
Natural environment	Limited in comparison to the scale of new investment, but objectively – a growing pressure on the natural environment, including the ecosystem of the Niepołomicka Forest. In spite of constant extension of the road infrastructure, the vehicle traffic is also becoming more and more burdensome.	In the scale of the CMA, negative impact of companies operating in the NIZ on the environment is slight.
Receipts from taxes (PIT, CIT, real estate)	–	–
Multiplier effects	–	Relatively low level of supply effects with respect to supplies of raw materials and production materials.
Social sphere	–	–
Transport	Absence of a motorway junction directly connecting the NIZ with the A4/ no possibilities for optimizing the road infrastructure in the commune/ liquidation of a railway line Podłęże-Niepołomice / growing vehicle traffic (supplier and passenger).	Absence of an effective, integrated group transport system in the metropolitan area with growing distances of work travels/ no solutions within the scope of the transport network in the scale of the CMA, including, in particular, in neighbouring communes (Cracow, Wieliczka).
Other technical infrastructure	First signs of absence of connection power with respect to energy infrastructure.	–

Source: own study.

operating in the NIZ (size, industry, outlet markets and other factors), in the recent years it was possible to observe strong resistance of the local economy to crisis.

Interpretation of mechanisms determining constant pro-investment policy does not provide an answer to the question about limits of growth. Undoubtedly, the NIZ is very necessary for the CMA and, in the prospect of several years – indispensable for development of metropolitan functions. No such attractive investment zone has been shaped so far in the CMA, which has at its disposal a similarly broad offer. In the course of time, it is necessary to expect that it will be more and more difficult to justify, to the commune inhabitants, the necessity of accepting new investors.

Further Extension of Operation in the NIZ – Local and Metropolitan Context

Development of the NIZ turned out to be favourable for the inhabitants of Niepołomice. Numerous new companies contributed not only to a drop in unemployment and increase in the standard of living of the inhabitants, but also, thanks to taxes which reached the commune budget – allowed for significant improvement of the quality of various municipal services. Development of the NIZ turned out to be important for the image of Niepołomice, including self-identification of inhabitants. Usually, they are proud of the success of their commune and strongly identify with it.

In Niepołomice, a clear 'growth coalition' has been shaped in the course of the years, within the scope of which activities are pursued to improve the quality of investment offers, in a significant degree via municipal investments, leading to the continual growth of real estate prices, including municipal real estate. In theory, such a layout has characteristic features described by Logan and Molotch (1987) as a growth machine. In the context of a slightly later concept of urban regime (Sagan 2000), such a layout may be interpreted as an instrumental coalition focused on constant development. It seems that currently the authorities of Niepołomice are, in these terms, an element of a broader coalition, whose objective is continual support and acceleration of development, further procurement of new investors, as well as – gradually, in a greater extent – new inhabitants. According to the ranking mentioned in the previous chapter, Niepołomice currently belongs to the most affluent communes in the region. The majority of municipal services (schools, roads, water, sewage systems, pavements, cleaning, public space, culture and recreation centres) are currently provided at a satisfactory level. In this situation, it is worth speculating whether further procurement of investors and, in consequence, increase in the commune's revenues will contribute to the significant improvement of the level and the quality of life of inhabitants. Response to the question about the viability of further extension of economic activity in the NIZ will, in the next few years, depend on the number of benefits that the local community gets from such development.

Interpretation of mechanisms determining the constant pro-investment policy of the town does not answer the question about the limits of growth. Undoubtedly, the NIZ is very necessary for the CMA, and, in the prospect of several years – almost indispensable for development of metropolitan functions. In the CMA, so far no such attractive investment zone has been shaped which would have a similarly broad offer at its disposal. However, in the course of time it is to be expected that it will be more and more difficult to justify to the inhabitants of the commune the necessity of accepting new investors.

Conclusions

Development of the NIZ influenced a number of social, economic, spatial and environmental processes in the commune. Intensity of this impact is a direct derivative of the number and the size of new companies which decided to commence and develop their activity there in comparison to the size and the potential of the commune. The commune, of a relatively small surface area, whose major part is occupied by environmentally protected areas, including the Niepołomicka Forest, with slightly over 20,000 inhabitants, accepted 60 new companies, mainly production plants where almost 5,000 people are employed. Particularly clear effects of this process were observed on the local labour market.

A definite majority of enterprises operating in the Niepołomice Investment Zone between 1994 and 2010 voiced their need for employees with vocational or primary education. Every significant stage of development by Niepołomice companies resulted in an increase in the demand for non-educated employees or employees with vocational education. From the point of view of the commune, it is necessary to consider it a beneficial phenomenon, which enabled, in a certain degree, the use of the commune inhabitants' qualifications. After 2002, the demand for more qualified production employees grew – with technical education and managerial personnel of middle and higher grades. These processes meant a clear quality change in the local labour market.

From 2007, the companies operating in the NIZ have had to face a difficult situation in the local labour market related to exhausting the possibilities of development on the basis of the existing pool of employees in the area of the county. One quarter of employees working in the industrial zone live in the commune of Niepołomice and at this level the possibilities of further growth in the number of commune inhabitants working for zone companies is significantly exhausted. However, from the point of view of commune inhabitants, the effects of this process were positive – in real terms, unemployment in the commune, especially among men, was reduced to the minimum level. Simultaneously, observing the volume of the commune's revenues from PIT – it is possible to notice, indirectly, a constant increase in the income of commune inhabitants. Companies operating in the NIZ also contributed to an increase in the commune's affluence, which is currently among the richest (per capita) in Małopolska Province. The most important sources of the communes self-generated revenues are: real estate tax, the commune's share in PIT and CIT. What is more, the image of the commune has changed radically – in the southern part of Poland, it is the synonym of success, wise management and skilful use of opportunities that are difficult to perceive by others. Development of the NIZ has also greatly influenced changes in the spatial management of the commune. In principle, all post-industrial areas that remained after the plants which operated in the period of People's Republic of Poland were liquidated were managed and transformed anew, acquiring new functions. The increase in the number of people in the commune results in a constant increase in the number of residential buildings, the great majority of which are single family houses, semi-detached houses and terraced houses. The establishment and development of the NIZ also forces the extension of the road system, which currently allows for the transfer of the majority of vehicular traffic generated by the NIZ outside the commune centre. However in total, the radical increase of traffic on local roads is one of the most burdensome effects of development of the NIZ for the inhabitants. Changes in spatial management also influence the natural environment of the commune. In particular, as a result of rapid urbanization, feeding grounds for birds and other animals living in the Niepołomicka Forest are getting smaller. Residential development occupies plots adjoining the forest, in many cases even at the very border of the forest.

On account of the relatively limited potential of Niepołomice, the effects of development of the NIZ started to leak outside of the area of the commune very quickly, that is by the end of the 1990s and this process has deepened in the subsequent years. After using the resources of the local labour market, companies operating in the NIZ started to use the broader resources of the entire CMA to a greater degree. What is more, services for business, sub-suppliers and customers operate in the area of Niepołomice to a very limited degree. Currently, the NIZ is definitely the largest and the most important investment zone of the CMA and Małopolska. Without doubt, more benefits go to entities located in communes other than Niepołomice within the borders of the CMA. Even though from the metropolitan perspective further development of the NIZ is necessary and important, an issue to which it is currently difficult to find an answer is the viability of further development of the NIZ from the prospect of inhabitants of Niepołomice. Hitherto extension of industrial and service functions in the commune only had a limited negative impact on the quality of life of inhabitants of Niepołomice. However, in the course of time, pressure continues to grow and benefits from development of the NIZ for the inhabitants are becoming less and less obvious.

References

Bank of Local Data from the Central Statistical Office. 2012. Available at: http://www.stat.gov.pl/bdl/app/strona.html?p_name=indeks [accessed: 30 April 2012].

Burbach, R., Nunez, O. and Kagarlitsky, B. 1997. *Globalization and its Discontents: The Rise of Postmodern Socialism*. London: Pluto.

Domański, B. 2004. Local and regional embededdness of foreign industrial investors in Poland. *Prace Geograficzne Jagiellonian University*, 117, 37–54.

Domański, B. 2010. Kapitał zagraniczny a rozwój polskich regionów, in *Przekształcenia struktur regionalnych – aspekty społeczne, ekonomiczne i przyrodnicze*, edited by S. Ciok and P. Migoń. Wrocław: Institute of Geography and Regional Development, Instytut Geografii i Rozwoju Regionalnego, Uniwersytet Wrocławski, 263–74.

Domański, B. and Gwosdz, K. (eds). 2005. *Dziesięć lat doświadczeń pierwszej polskiej specjalnej strefy ekonomicznej. Mielec 1995–2005*, Kraków-Mielec: Institute of Geography and Spatial Management UJ and ARP.

Grabher, G. 1994. The disembedded regional economy: the transformation of East German industrial complexes into western enclaves, in *Globalization, Institutions, and Regional Development in Europe*, edited by A. Amin and N. Thrift. Oxford: Oxford University Press, 177–95.

Hardy, J. 1998. Cathedrals in the desert? Transnationals, corporate strategy and locality in Wrocław. *Regional Studies*, 32(7), 639–52.

Jarczewski, W., Huculak, M. (eds) 2011. *Sukces polityki proinwestycyjnej. Niepołomice 1990–2010. Wnioski dla władz lokalnych*, Kraków: Instytut Rozwoju Miast.

Logan, J. and Molotch, H. 1987. *Urban Fortunes: The Political Economy of Place*. Berkeley: University of California Press.

Markowski, T. and Marszał, T. 2006. *Metropolie. Obszary metropolitalne. Metropolizacja*. Warszawa: KPZK PAN.

Ocipka, J. 1999. Specjalne strefy ekonomiczne w Polsce. *Ekonomika i Organizacja Przedsiębiorstwa*, 2, 18–20.

Oinas, P. 1997. On the socio-spatial embeddedness of bussines firms. *Erdkunde*, 51.

Oinas, P. 1998, *The Embedded Firm? Prelude for a Revived Geography of Enterprise*. Helsinki: Helsinki School of Economics and Business Administration.

Ranking gmin małopolskich MISTIA 2012. Available at: http://www.mistia.org.pl/pokaz_strone.php?s=635 [accessed: 15 April 2013].

Stryjakiewicz, T. (ed.) 2004. *Wpływ inwestorów zagranicznych na rozwój regionalny i lokalny na przykładzie GlaxoSmithKline Pharmaceuticals S.A. w Poznaniu*. Poznań: Bogucki Wydawnictwo Naukowe.

Sagan, I. 2000. *Miasto. Scena konfliktów i współpracy. Rozwój miast w świetle koncepcji reżimu miejskiego*. Gdańsk: Wydawnictwo Uniwersytetu Gdańskiego.

The Word's Best Brands. The success stories – and the failures – of this year's Top 100. 2007. *Bussiness Week*, 7 August 2006.

PART III
Innovative Small Firms – Strategies and Behaviour

Chapter 9
Beyond a Creative Class Hotspot: Innovation in Independent and Corporate Coffee Shop Enterprises

Mathew Novak

Creativity in the Coffee Shop

A palpable buzz fills the air inside cafés and coffee shops.[1] Artists, designers, programmers, and others in the creative industries work away at their laptops, drawing inspiration from not only the caffeinated beverages, but the spaces themselves. Being in a coffee shop stimulates all five senses, thus promoting the creative process. The coffee shop as an engaging environment is not, however, a new phenomenon. The West's first coffee shop opened in Oxford, England in 1650, and the concept spread to the other intellectual and influential cities of Europe (Pendergrast 2010: 12). They were colloquially known as "penny universities" since the cost of tuition to take part in the inspired theoretical and political discourses was the price of a cup of coffee.

Beyond the coffee shop as both an incubator and a signifier of creativity, the coffee shop enterprises themselves are often highly innovative businesses. Coffee shops employ new business practices to expand their markets, locating in once undesirable areas like gentrifying neighbourhoods, and entering new suburban and rural markets. Through creative marketing coffee shops further expand their reach by drawing in new demographics. Additional customers are attracted by new product offerings and delivery methods. The arrival of gourmet coffee in the late 1970s was a major innovation in the consumer coffee market (Roseberry 1996), a market which has continued to evolve over time.

The following chapter looks at the numerous innovations that the coffee industry has undertaken since the advent of the gourmet coffee trend in the late 1970s. Innovations are found in the location of shops, moving beyond the traditional urban core, and entering new economic spaces. Others are identified in the marketing strategies aimed at new demographics, including niche segments

---

1 Café and coffee shop are used interchangeably in this chapter, both referring to a place whose primary offering is drip coffee and espresso based drinks, but which may also serve other beverages such as teas and juices, and offer a limited range of food, especially pastries.

of the market. Further innovations are shown in the technologies employed in making coffee, and delivery formats, namely the arrival of the drive-thru at the coffee shop. In discussing these various strategies, the resourcefulness of small independent coffee shops, as well as their relative advantages and disadvantages, are acknowledged as they compete with their large rivals for customers.

The small shops have several benefits over their large rivals, including the ability to enter niche markets and to quickly adapt to change. It will be argued that the small scale of the independent enterprises allows for more flexibility in their market positioning and ability to innovate since the large corporate inertia is not present. Furthermore, since these two types of enterprises exist in largely different markets, their relationship is less combative, and could even be seen as parasitic with the independents feeding from the market created by the corporate chains, much of the time without noticeably harming the large players. In a time when corporatization is the dominant trend in the retail and service sectors, it is notable that small, independent coffee shops may compete with, and actually benefit from the large players.

Study Area and Methodologies

As the birthplace of Starbucks, the world's preeminent corporate coffee chain, the State of Washington makes for an ideal and illustrative case study of innovation in coffee shop enterprises. Starbuck's head office remains in Seattle, the state's major metropolitan area. But the region is also a fertile bed of coffee innovation beyond the Starbucks' boardroom. There are hundreds of independent coffee shops dotting the urban and rural landscapes of Washington. The interaction of the independent and small chain coffee shops with the global behemoth that is Starbucks provides illustrative examples of how innovation occurs in various scales of enterprise

The Washington State Department of Revenue provided a database listing all incorporated businesses in the state. Coffee shops, independent and corporate, were selected for by querying the name of the business for iterations of Coffee, Café, Java, Latte, and Espresso. Although many other entities sell coffee, only those whose primary product is coffee, as indicated by their name, were used in this analysis. Individual records included addresses, allowing for the points to be spatialized using GIS. Maps made from this data show the location of coffee shops across the state; of note is their distribution in rural versus urban areas.[2]

After mapping the macro locations of coffee shops across the state, field observations were conducted to gain a better understanding of their operating practices. These observations reveal the characteristics of the coffee shops, their customers, and how they fit within the local fabric. Analysis focused on the form and function of coffee shops in the central Washington communities of Yakima, Ellensburg, Wenatchee and Leavenworth, and numerous smaller towns and

2 For this research urban areas are defined as census places with a population of at least 10,000 in 2010.

rural areas throughout the region. The size of the building housing the shop, its furnishings, advertisements, marketing programs, and menu offerings were all documented. These observations allow for a detailed comparison of business practices between the independent and corporate shops. Further information about business practices was gleaned from select coffee shop websites and the existing literature on the practices of Starbucks. These sources provide insight into coffee enterprises as they innovate to compete for new customers.

The Culture of Coffee Shops

Coffee shops are settings of social theatre, exemplifying Oldenburg's (1999) notion of a "third place", that is, somewhere other than home and work where people mingle in the social realm. Increasingly many patrons use coffee shops as office and creative space, in addition to social space (Bryant 2009). The majority of shops offer Wi-Fi connections, allowing customers to download and disseminate information over the internet, creating connections beyond the walls of the coffee shop.

Many authors and researchers recognize the association between coffee shops and creativity, and often use their presence as a sign of an area's creative potential. In his research on the "creative class", Florida (2002) discusses how coffee shops are where creative people congregate for both work and leisure. Florida points out that if an area houses a large number of cultural and entertainment amenities, including cafés, restaurants, and art galleries, there will likely be found a high concentration of individuals that comprise the creative class, that is artists, new media professionals, computer programmers, and other workers in the growing, high-order service sector.

The social commentator David Brooks (2000) describes "latte towns" as places where high numbers of "bourgeois bohemians" (bobos) are likely to be found living and working. Bobos, like Florida's creative class, are generally highly educated, employed in knowledge-based occupations in the post-industrial economy, and frequently haunt coffee shops in both their leisure and work times. Latte towns are commonly located near a major college or university campus since many professors, researchers, students, and others who work and play on campuses epitomize and foster the bobo culture. In his study on American college towns, Gumprecht (2008) frequently mentions coffee shops as a typical business found in trendy commercial districts in these distinctive locales. College towns, with their students, professors, artists, and consultants, are a hotbed of creative activities, where one characteristically finds café culture.

Coffee shops are often a sign of gentrification, the encroachment of trendy stores, well-paid jobs, and creative class demographics, into existing, usually poor, inner-city neighbourhoods, typically displacing the existing retailers and households (Lees, Slater and Wyly 2008). Coffee shops are among the retailers and services such as art galleries, bistros, and antiques stores that enter urban spaces in transition. Since such processes raise rents and change the existing dynamic of the area, they are often seen as negative by the academics who study them, especially when involving corporate chains such as Starbucks who dislocate traditional retailers (Smith and Williams 1986). Despite the negative implications, the entry of gentrifiers, such as cafés, into new urban districts is a harbinger of the emerging creative potential of the area, and entails innovations in business practices by uprooting the traditional spatial structure of the city.

Innovation in the Retail and Service Sector

Players in the retail and service sectors are persistent innovators. Within a perpetually changing market, outlets from fast food restaurants to high end department stores must be innovative in order to continue attracting customers, thus maintaining profitability. Retailers and service enterprises that fail to innovate will likely not last long in this highly competitive and dynamic arena wherein both the supply and demand sides of the market are prone to incessant change. Customers' shopping preferences, tastes, and resources must be accounted for, along with their demand for new products and novel experiences. On the supply side, competition is brisk between stores. Each new innovation brought on by a competitor compels the store or service to match or better it. Further supply dynamics involve new products, commodity chain negotiations, and logistics.

Numerous examples of innovative practices abound in the retail and service sectors. Much of Wal-Mart's phenomenal success is attributed to its innovation in restructuring the supplier–retailer relationship (Bloom and Perry 2001, Fishman 2006). McDonald's standardized the production of its food products much like an assembly line to ensure quick and dependable service, and replicated these practices across its franchises (Noren 1990). Zara and H&M offer chic-cheap clothing to the middle classes, and are able to stay at the front of fashion trends due to their ability to bring new fashions to market quickly (Ferdows et al. 2003, Hilletofth and Hilmola 2008). Starbucks brought the Italian coffee bar first to America, and then the globe, and in so doing was an important pioneer in the gourmet coffee industry (Schultz 1997).

All of these examples involve large corporations bringing innovations to the market through their chains of outlets in order to increase their competitiveness and profitability. The chains have been very successful in their ability to innovate, and along with the advantages realized by economies of scale, have dominated many segments at the expense of the traditional small businesses (Jones and Simmons 1990: 76–8). Increasingly 'ma and pa' shops are relegated to the margins in many segments of the market. Many lament this loss, as it means a homogenization in the retail landscape across countries and around the globe (for example Wurgaft 2003). Today the same shops can be found on an American main street as along a British high street, and even in a new Chinese shopping mall.

Innovations in Location

A common way that retailers and services employ innovation in their business strategies is by choosing new locations for their outlets. Businesses choose a location in order to be as attractive as possible to their existing and potential customers, thus increasing their profits by selling more goods. These choices become engrained, creating a typical spatial retail structure in cities, such as was found in the traditional downtown shopping districts that dominated most cities until the mid-twentieth century (Berry 1963). Spatial innovations occur when retailers choose to disrupt the existing spatial structure of the city by moving into new areas. Similarly, spatial innovations also occur when retailers enter new economic spaces previously devoid of retailers, such as in marginal spaces in the existing retail structure, or entering into partnerships with institutions.

The most prominent example of innovations in business locations over the last century is the switch from the city centre to suburban shopping in towns and cities across North America. The traditional downtown shopping district began to decentralize in the late-nineteenth century (Conzen and Conzen 1979, Novak and Gilliland 2011). Before then the spatial structure of cities was quite simple, with most retail activities taking place in the core. With high competition and accompanying land values, some shops began decentralizing, taking on the risk of disrupting existing spatial patterns and relocating on their own.

These new spatial patterns in the retail landscape were accelerated in the second half of the twentieth century as families began moving into single family homes in the suburbs. The traditional downtown core became less attractive since customers moved further away, and the rise of the automobile led to congestion on the streets and parking was scarce. A radically different shopping environment emerged in the shopping malls and retail plazas that were developed primarily in the expanding suburbs in practically every market during the second half of the twentieth century. The new paradigm in where and how people shopped is defined by shopping centre's controlled environments; climate, public access, and store selection are all under common management.

Department stores are among the most notable instances of stores leaving their traditional downtown locales, and opening branch stores in the suburbs (Longstreth 1997). Filene's in Boston, Hudson's in Detroit, Marshall-Field's in Chicago, and Bullock's in Los Angeles all began opening branch stores in the suburbs in the mid-twentieth century. Doing so allowed them to expand their market by catering to the new suburbanizing populations, and escape the high rents and competition of the core.

Cafés and coffee shops followed a different locational trajectory from most other stores and services, resulting in distinctive contemporary spatial patterns. Traditionally, like most retailers, coffee shops were found in the central areas of cities; however, they remained exclusively in the core throughout much of the twentieth century, even as many other sectors began relocating outside the core. Whereas most other retailers followed their customers out to the suburbs, and entered new shopping centre developments, coffee shops persisted in the traditional shopping districts.

The arrival of the gourmet coffee trend in the 1970s began in city centres and trendy urban neighbourhoods where their typical clientele – hip, yuppie, young, wealthy, artistic –was found (Roseberry 1996). For example, this demographic is typical in Berkeley California, where Peet's Coffee, a forerunner in the gourmet coffee industry, was founded in 1966. Even Starbucks, which today is as common in suburbia as in traditional urban areas remained exclusively within the traditional spatial mould – downtowns business districts and trendy urban neighbourhoods – throughout much of its early expansions. Moreover, until recently many small towns and rural areas were devoid of cafés. Baring some college towns and other enclaves, coffee shops remained a quintessentially urban amenity. Thus coffee shops occupied a small market area, and as such catered to a relatively limited demographic.

Downtown fortunes continued to wane throughout the late twentieth century in many American cities, while the suburbs and exurbs boomed. Thus impetus was placed on coffee shops to look for expansion into these new realms. The push to decentralize was accompanied by market saturation in the existing areas, after extensive expansion by corporate and independent shops alike. Furthermore, coffee began entering new demographics, as middle and even lower income groups began taking up the habit of gourmet coffee. These groups were less likely to haunt the city cores, so cafés followed them to where they lived, worked, and shopped.

The growth of specialty coffees fuelled much of the innovations in coffee shop locations. Both corporate and independent shops had exhausted the traditional markets, and the real expansion was in moving towards the mass market. This market expansion necessitated making the outlets convenient to the mass market, which had largely left the city for suburbia. As a result, coffee shops began moving to the suburbs and exurbs, setting up in shopping malls, and along automobile dominated commercial strips.

Examples of businesses following their customers into new areas is not particularly novel. Such patterns have happened repeatedly in nearly every sector of the retail and service industries. Department stores left lower Manhattan in the nineteenth and twentieth centuries, following their wealthy customers into newly fashionable midtown locations (Domosh 1990). The innovation lies in where coffee shops opened up in these new realms. Traditional standalone cafés in downtowns and along retail strips are joined by cafés in shopping centres, inside public and private buildings, at freeway exits, within parking lots, and numerous other novel locations.

Today, coffee shops are prolific in Washington (Figure 9.1). In 2010 there were 2072 outlets whose primary product was coffee, as indicated by their name. That equates to roughly one store for every 3100 citizens. Although the majority of coffee shops (75.9 per cent) are in urban areas, there are still 500 in small town and rural districts (Figure 9.1). Thus, coffee shops are no longer urban phenomenon, with the number of stores per capita roughly equal between rural and urban areas. Coffee shops have become a ubiquitous part of Washington's landscape.

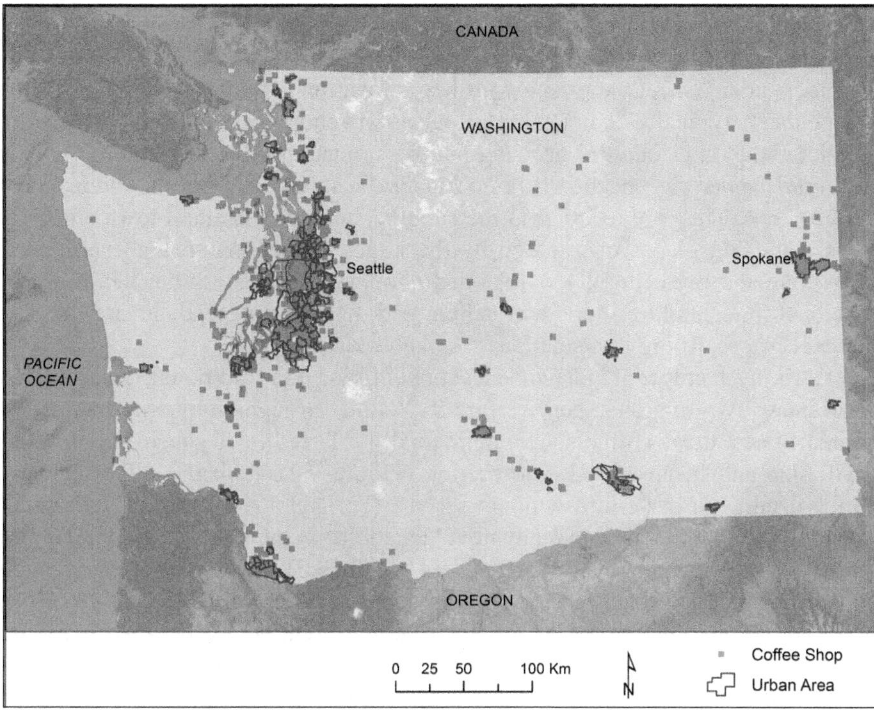

Figure 9.1 Locations of coffee shops in Washington State
Source: own map based on data from Washington State Dept. of Revenue (2010).

One finds coffee shops in all manners of public and private realms, inserted wherever a captive customer base might appear. Airports, office towers, museums, military bases, and college campuses all commonly have at least one coffee shop, and at times can house numerous. Coffee shops have entered these non-traditional spaces as the demand for coffee consumed outside the home rises. They are integrating with existing businesses and other entities, usually paying high rents for the promise of a captive market without other competitors.

A notable example of this entwining of coffee shops within larger institutions are the kiosks found inside many large stores selling a full complement of coffee drinks. Throughout Washington it is not uncommon for grocery stores, gas stations, and department stores to have a coffee kiosk. Most of the host stores are corporate chains, as are the coffee shops. Instances of independent stores hosting cafés also are also present, including fruit stands, gift shops, and book stores. In these independent stores one normally finds non-chain coffee operations. Both the corporate and independent kiosks are smaller than typical standalone outlets, but offer the same drinks and other menu items under the recognized branding. Although most have a few seats for customers, there is not the atmosphere associated with the classic coffee house. Coffee is primarily ordered to go, consumed while shopping for groceries or browsing the book shelves.

Independent coffee shops show further innovation in their locations. While the independents typically do not enter the successful shopping centre and chain store locations, they are found in growing numbers outside of the urban core. Artistic and trendy inner city locations are no longer the sole domain of the independent coffee shops. They are locating in the suburbs, and even in small town and rural areas, choosing places overlooked by the larger chains. Whereas the corporate shops like the predictability of planned shopping centres, or the flexibility of spacious, free-standing shops with ample parking, the independents are locating in novel places, fitting into small, overlooked spaces.

A striking example of independent shops entering non-traditional, underutilized spaces are Washington's *coffee shacks*. Found throughout the north-western United States, these coffee shops, colloquially referred to as shacks due to their small size and insubstantial construction (Figure 9.2), typically offer specialty coffee drinks via drive-thru window service (Frenkel 1995). Due to their small size they are able to locate in many novel locations: in the boulevards along major streets, under signs, and inside parking lots (Figure 9.3). These spaces are not large enough for most other enterprises, but are ideal for the small coffee shops. Since they take up what is otherwise unused space, their rent is low, but their exposure is high locating along busy corridors. Many of the shacks are so small that they are portable, and can move if an existing place does not work out due to either few customers or problems with the rental agreement. Thus, these coffee shacks are not only locating in innovative spaces, but are also innovative in their flexibility to relocate in response to market pressures.

In moving to the suburbs, exurbs, small towns, and even rural areas, coffee shops are expanding their markets. By breaking the mould of the traditional downtown coffee shop, these spatial innovations cater to new clientele, thus growing markets. In locating in spaces that their corporate competitors are unable or unwilling to enter, the independent shops are finding much success, despite their limited resources.

Innovations in Store Format and Product Delivery

New locations for coffee shops are accompanied by innovations in the format of the shops, as well as methods of product delivery. The traditional store layout found in city centres – namely long, narrow spaces with a walk-up bar on one side and tables along the other and large windows facing the street – have given way to a variety of layouts and designs. Likewise, no longer are the stores solely accessed by pedestrians strolling down the sidewalks along the shopping streets. Now coffee shops accommodate the car-bound customer with large parking areas, and many offer drive-thru window service so that customers do not need to leave their automobile to get their caffeine fix.

Figure 9.2 A coffee shack near Yakima, WA. Note the unsubstantial nature of the structure

Source: author's photo.

Figure 9.3 A coffee shack in Wenatchee, WA inserted in a small, otherwise unused part of a store's parking lot

Source: author's photo.

Some of the new coffee shop formats and delivery methods are a reflection of necessarily doing things differently due to the different spaces in which the shop is found. Others are due to new ways of conducting business to increase efficiencies, or to attract new customers. The aforementioned kiosks and coffee shacks are examples of new formats of shops, and offer new ways of delivering their product.

The old model of a café where patrons ordered drinks at the espresso bar and then consumed them at the tables that lined the opposite wall has been replaced by coffee shops as stand-alone kiosks housed within larger buildings. Hypermarkets, airports, universities and other public and private institutions now often have coffee kiosks, which are much smaller than the typical coffee shop, and usually have only limited seating. Offering coffee in such situations has logistic issues that need to be addressed, notably the lack of a separate kitchen and storage space; the entire operation is observable to the public. Also, the logistics of deliveries and refuse collection are different since the shops do not have their own access, but are imbedded within their hosts.

Coffee shop operators have devised strategies to overcome these limitations in operating stand-alone kiosks. Most notably the corporate chains which are ubiquitous operators of these kiosks in the chain stores, airports, and other major institutions, have compartmentalized and standardized the production of coffee in their shops, so that small spaces without prep areas are sufficient. Disposable cups eliminate the need for dishes and dishwashers. Food is delivered, rather than made on site.

Both sides are often eager to enter such partnerships since they can be highly lucrative. Coffee shops are able to expand their market by entering these new economic spaces. The host stores and institutions are also agreeable since they can charge rents for spaces, and may even negotiate for a cut of the profits from the outlet. The hosts also benefit since the kiosk makes their outlet more desirable, drawing in customers for a coffee and then picking up milk, or making their wait at the airport more enjoyable.

The coffee shack is similar to the kiosk in that it is much smaller than the traditional coffee shop, and disrupts the traditional service model. Whereas the kiosk is typically part of a corporate chain operating inside other larger institutions, the coffee shack is usually an independent operation, or part of a small chain, and occupies a freestanding site. Being small, and of unsubstantial construction means that the coffee shacks are open to a much larger number of operators, since capital requirements are limited.

Most of the coffee shacks do not offer traditional walk-in service; the premises are simply too small. Their business model is based on serving automobile-bound customers via a drive-thru window. Adapted from its widespread use in the fast-food industry, the drive-thru window is becoming part of coffee culture. Beyond the coffee shacks, many other coffee shops, including Starbucks, are installing drive-thrus to allow customers to remain in their car when ordering and consuming their coffees and espressos. Although walk-ins remain a common service practice, drive-thrus are becoming widespread, allowing coffee orders to be placed from the comfort of one's private automobile.

The application of drive-thrus in the delivery of coffee opens up the car-bound market, which is large in many Western countries. In the United States, 91 per cent of households own an automobile, and most have multiple (United States Bureau of the Census 2010). Catering to this large demographic spurs sales, since many like the convenience and speed of the drive-thru window. Such practices, however, disrupt the traditional notions of coffee shops as "third places" (Oldenburg 1999). Although making the coffee more accessible, the drive-thru window also isolates the customers from the creative buzz of the coffee shop.

These new formats of coffee shops, and the delivery methods they utilize, demonstrate how the coffee industry is innovating in order to grow. Coffee is entering new economic spaces, while simultaneously breaking the mould of the traditional coffee shop. In so doing coffee shops are becoming more accessible to their customers. More importantly, however, they are expanding beyond their traditional customer base, and entering new market segments. By moving to the suburbs, and offering drive-thru service, they are decidedly entering middle class markets, away from their traditional niche segments found in the urban cores. In the following section we further examine how coffee shops are moving into new market segments by analysing their creative marketing practices.

Creative Marketing to New Audiences

Starbucks is notable among major retailers and services for having a limited marketing budget. This in itself is innovative, since most other large corporations, both in the food services specifically, as well as consumer goods in general, typically spend lavishly to market their products. Companies as diverse as Nike, Target, McDonald's, Apple, and Ford all spend vast sums annually on advertising. Starbucks only introduced television advertisements after severe financial problems in the mid-1990s and still spends relatively little on marketing in comparison to others in the retail and service sectors (Bryant 2009: 40). That they were able to stimulate demand for coffee, as well as foster a highly recognizable and fashionable name brand with relatively little spent on advertisements, is truly innovative.

Part of Starbuck's success was identifying untapped market demand and its ability to enter and establish itself at the beginning of the specialty coffee market. Their original outlets in Seattle were among the first gourmet coffee purveyors in North America.[3] Another key to Starbuck's success is through the building of attractive outlets in desirable areas. Starbucks uses the atmosphere of its stores to heighten the desirability of the brand (Bryant 2009). As a result, Starbucks is not just selling a product, but also an image epitomized in the spaces it occupies. This innovative strategy has been employed elsewhere, but seldom as successfully as in the Starbucks model.

---

3 Peet's Coffee, the other major early roaster, was based in San Francisco. Elsewhere in the country gourmet coffees were nearly non–existent.

In creating atmospheric stores, and a desirable brand image, Starbucks is able to offer affordable luxury to its customers. Although four dollars is a lot to spend in comparison to the traditional mass-market coffee, it is attainable by most of the population, especially if done only occasionally. Thus Starbucks attracts segments beyond its original urban, wealthy, image conscious, coffee connoisseurs, moving into decidedly middle class segments.

While Starbucks and the other corporate chains foster consistency and coherence to an overarching brand identity, the independent shops are free to adopt a wide range of marketing devices to sell their product. Some adopt lower prices, and selling in greater quantities to generate profits. Others may enter niches in the market to attract a specific demographic, differentiating themselves from their corporate as well as other independent competitors.

With homogenization in the urban retail landscape resulting from corporatization, simply being an independent coffee shop can be advantageous in attracting certain customer groups. Many disavow the corporate model, and prefer to frequent local, independent shops (Thompson and Arsel 2004). These shops differentiate themselves from their corporate competitors through funky design, employing edgy baristas, playing experimental music, hanging provocative art on the walls, and offering events such as poetry readings and open mic nights; thus catering to the desires of their target customers. Simply being independent makes many small shops able to compete in some markets with their corporate rivals.

Other independent coffee shops come up with creative marketing that orientates them along a theme. They devise names, advertisements, and structures that follow a common image. A notable example is HeBrews, a coffee shack operating in Yakima, Washington (Figure 9.4). Its name prominently displayed on the building, a biblical pun, is accompanied by a reference to a verse from the scripture. Most apparent is its design, taking on the appearance of a small chapel, painted in white, complete with steeple. If only for the novelty, HeBrews may be successful, and their owners cannot be accused of being unoriginal.

Perhaps the most striking examples of creative marketing adopted by some of the coffee retailers employs the oldest device in the book – sex. Across Washington one finds coffee shops where scantily clad women prepare the drinks. In employing 'bikini baristas', these coffee shops cater to a niche market by offering overt sexualization to sell coffee. Provocative names such as "Moulin Brew", "Java Jigglers", and "Dream Girls Espresso" are used by the owners to advertise to the clientele what the coffee shop offers in addition to the coffee itself. Some even go so far as to charge extra for a photograph with the baristas, and have theme days where the baristas dress up in western wear or as school girls (Cow Girls Espresso 2013).

Obviously such "bikini barista' establishments are only targeting a small subset of the market. With marketing campaigns considered offensive to some, these shops are decidedly not positioning themselves for the mass market. They are successful by differentiating themselves from their competitors, and actively targeting a niche market. Cow Girls, a local independent chain, has 14 outlets in the Seattle area. The once high-society café has met the low-market of the sex industry.

Figure 9.4 HeBrews in Yakima, WA utilizes distinctive and clever marketing to attract customers
Source: author's photo.

Flexibility, Innovation, and Scale of Enterprise

The scale of enterprise enables and dissuades different forms of innovation in the marketplace. Small, independent coffee shops do not have the capital or human resources of their large competitors, and thus are not able to conduct extensive market analysis, roll out coordinated advertising, or outfit luxuriously appointed shops in desirable shopping areas. Yet their small size also enables them to be innovative, since they do not have the inertia of large bureaucracies and infrastructures, and are able to diversify to fit small segments of the market.

The previous example of the bikini baristas and the use of unconcealed sexualization to sell coffee illustrates how enterprise size plays a role in innovation. It is only the small independent coffee shops that can target a niche market, since in catering to a small fraction of the overall market, they are driving away many other customers. In employing bikini baristas, these shops have marginalized themselves from the mass market. Their overt sexualization of the experience is not to most customers' tastes, who will not patronise such a shop.

Large chains such as Starbucks must align themselves with as broad of a market as possible. Doing so ensures that profits are maximized throughout an extensive network of stores. Furthermore, they have an image to uphold, and cannot risk tarnishing it with negative associations. Independents do not have the same extensive network to keep going, nor a valuable brand image that needs to be maintained, so they can target themselves at specific niches, even ones that drive away the typical mass market. The small shops that choose an exclusionary niche, such as the bikini baristas, may be successful since they are serving a small, but consistent segment, and profit from the novelty of the experience which cannot be found elsewhere. Such niche players are securely protected from corporate competition since the exclusive yet exclusionary market disavows larger competitors from entering.

Starbucks and other large corporations are prevented from entering some niche markets, namely those that are associated with negatives. Their sheer size necessitates a business model targeted at the mass market, where most consumers and thus profits are found. Starbucks has increasingly been moving towards the middle as it opens more shops. Originally, Starbucks was situated in desirable metropolitan areas, and a brand associated with an exclusive, urban clientele. Growth has necessitated moving beyond its original upper class market. Starbucks has securely entered the middle class as it opens coffee shops in the suburbs and offers drive-thru service.

In targeting the mass market, large corporations have resources that the small independents do not. They have financial and human resources to run large marketing campaigns, spending lavishly on making a recognizable brand image. Throughout much of its expansions Starbucks did not advertise when it was an exclusive brand, but as it moved into the mass market it began to use print, television, and other media advertisements to draw in consumers and solidify its image. Large chains also have the human resources to run an effective and targeted marketing campaign through a creative team of marketers who collect extensive information about customers, and gear their messages accordingly. They use sophisticated technologies, such as GIS, to locate new stores and expand spatial coverage (Hernandez 2007). Once they decide on an area, they have the resources to pay high rents, and outfit elaborate stores.

In addition to their large marketing resources, the corporate coffee shops have the ability to promote technological innovations that would be unavailable to smaller-scale players. Starbucks introduced the Verismo espresso machines in the first years of the new millennium, reducing the need for experienced baristas through standardization and automation. At the time they were heralded as revolutionary, although recently they were replaced with a more traditional machine that brought back some of the atmosphere to the coffee shop (Bryant 2009: 46). Starbucks is currently experimenting with new ways of payment, namely through smartphones. In 2012 it invested $25 million in Square, a technology firm that pioneered methods for wireless payment through smartphones without the need for physical currency (Efrati and Gasparro 2012).

Obviously the small coffee shops do not have the financial resources to purchase tech firms, or commission their own line of espresso machines. They must be content with obtaining the technology once it filters down to a level where they can adopt it. Thus the small players lag behind in many innovations, especially those that require large amounts of capital to fund. The independents do, however, have other innovative advantages over their large counterparts – namely their smallness allows for innovations to be quickly adopted. Whereas large companies have an accompanying large bureaucracy that innovations must work their way through, small entities can quickly innovate without the inertia of numerous strategic meetings, working groups, and approvals through which the new ideas must travel.

The small chains also have the benefit of being imbedded in the local market. Most independent owners live in the same market area that their shop serves, allowing them to continue to monitor local market conditions. This first-hand knowledge, along with connections to local businesses, allows the independent coffee shops to quickly respond to local conditions and market demands. Corporate chains have head offices away from the actual operations; often thousands of miles and even countries separate where the decisions are made from where they are implemented. Thus, they lack local knowledge of the market dynamics of a given location beyond what their analytical divisions can gather from surveys, the census, and other metrics. When a local major employer closes, for example, the independent coffee shop owner typically will know in advance, and can adjust his or her pricing structures and product offerings to reflect this change in the market. Corporate chains, meanwhile, do not have the local knowledge, and must react much later, and often more rigidly, in line with corporate policies.

Further obduracy in the corporate infrastructure is due to their need to uphold the brand image. Corporate shops are reliant upon the desirable brand image they have fostered over time to attract customers. They are less likely to innovate, especially with radical ideas that have large reward, but also large risk, since they might tarnish the brand if failure occurs. Small shops rely less on brand recognition, thus spurring innovation by reducing the cost if the new idea is unsuccessful in the marketplace.

Low Entry Barrier Facilitates Creativity and Innovation

A low entry barrier, measured in terms of both financial and human capital, allows for innovations to flourish in the coffee industry. Whereas many businesses start-ups require substantial capital investments, and entail finding and training numerous skilled employees, coffee shops can be started with little capital to build and outfit the establishment, and can be run with only a handful of employees. With low entry barriers, more players may come to the market, bringing their own innovations. Furthermore, since financial investments are small, there may be more innovation since the outcomes of failure are minimal.

Most coffee shops are operated with fewer than 10 employees, and many times have less than five. In some cases, such as the coffee shacks, the owner may also be the sole employee. Employees do not necessarily need any experience in a coffee shop environment as they can be trained on the job. Advances in equipment mean that the machine does much of the work at the press of a button – grinding the coffee, measuring it and tamping it into the portafilter, running through the correct amount of water at the correct temperature, and disposing of the grounds. Some machines even have automatic steamers for the milk. Although now returning to more manual systems, Starbucks used such automated machines in their stores for many years, mass producing predictable coffee, and reducing the need for skilled baristas.

While there is truly a lot of skill in the art of making great espresso and its derivative drinks, many of the shops do not sell a premium product that relies on quality coffee beans and brewing techniques. Rather, they are satisfying customers who want drinks in which the coffee is secondary to the other components, including flavoured syrups, sugars, heavy and whipped creams, and at times actual pieces of chocolate, caramel, or cookies, making the drinks appealing to sweet and rich tastes. Whereas a great cappuccino relies on quality in its simplicity, few stores sell these in significant numbers; rather, consumers are drinking highly flavoured, elaborately concocted drinks. With the quality of the coffee being masked, baristas no longer need to master the skills of espresso brewing, and milk frothing. For many of the drinks all that is required is that the barista follow basic recipes with the correct formulas of syrups added to shots of poor quality espresso and improperly steamed milk.

Although specific financial data is not available, it is reasonable to assume that capital investment for the building and chattels is limited. 'Coffee shacks' is an appropriate moniker for these establishments; many are less than 10 square metres in size, and built of insubstantial materials. Those shops that rent space have no capital investment in the physical structure itself. The equipment required to outfit the shops is not extensive. Although professional grade espresso machines and grinders can cost as much as some of the coffee shacks themselves, with some models running over $20 000, many are less than half that cost and constitute the main equipment expense. Other equipment needs are minimal, such as blenders and small refrigerators.

In comparison to other food services, coffee shops require far less human and financial capital than full-scale restaurants, or even quick service delis – both of which require an entire kitchen to be fitted, including stoves, freezers, pots and pans, dishware and flatware. In serving a limited product line, coffee shops reduce the amount of equipment needed. Also the staff requires less training since they need master only a limited range of options. Using only disposable cups, coffee shops can eliminate the initial investment in dishware, as well as the need for commercial dishwashers and their operators. The buildings themselves are typically smaller; in the case of the coffee shacks costing a fraction of other food service establishments to build or rent. In terms of human capital, restaurants require cooks who take years of training, and servers who need excellent interpersonal skills. Coffee shop employees typically spend less time interacting with their customers and are required to produce a limited range of products.

The high entry barrier of most other businesses eliminates many potential players from the marketplace. The lack of financial resources precludes many from opening a business, even if they would be successful by offering an innovative new experience for customers, or better business management practices. Coffee shops, with their low capital investment, are a sector available to many more players. With more players in the arena, innovations are fostered as each brings creative ideas to the market.

The limited human capital needed in terms of both number of employees and their requisite skills also fosters coffee shop innovations. Business owners can open a shop even if they have little or no existing knowledge about the coffee industry as they can quickly learn the skills needed to concoct the various beverages on offer. Coffee shops may enter markets with limited labour pools since they do not need to hire large numbers of skilled workers. The arrival of coffee shops in exurban, small town, and even rural areas is made possible by the limited need for skilled employees.

A high entry barrier also makes conservative business practices, since the investors do not wish to risk their large sums of money on risky prospects. Thus, those in businesses with large start-up investments are likely to choose to enter safer markets and employ proven marketing strategies. Innovations often entail risk as the novel is unproven and the prospects for failure are high. The lower capital investments in coffee shops result in a higher tolerance for risk. Coffee shop owners and their investors, if present, are less averse to trying new business practices since their initial investments are so much lower. If a coffee shop opens with an innovative new format or location, and is ultimately proven not to work by the market, the owners and investors are out of relatively little capital.

The Relationship between Corporate and Independent Outlets

In some ways the small coffee shops may actually benefit from the presence of the large corporate players. Corporate chains have conditioned large segments of the market to drink gourmet coffee outside of the home on a regular basis. For some consumers take-away gourmet coffee is now a daily habit, which can become quite expensive at an average of $4.00 per cup, while for others it remains an infrequent indulgence. Starbucks and other purveyors have made coffee an affordable luxury, fuelling the gourmet coffee trend, and instilling in the public a desire for elaborately concocted espresso-based drinks. Arguably the independent shops would not have the reach to introduce and condition the market to consume low-fat, venti, caramel lattes made with soy milk and three shots of espresso.

Due to their limited resources, and insubstantial reach, the independent coffee shops do not have the ability to shape the market like the large players do. Rather, the small coffee shops have benefited from the large players stimulating new markets for their product, and moved in to offer it in innovative ways. The independents build small shops in underutilized locations, enter new markets that their competitors avoid, and apply clever marketing such as we saw with the theme shops or the bikini baristas.

The large and small enterprises are many times working in distinct markets. Starbucks and others have defined market parameters, wherein they will only locate in specific areas, and target certain demographics. Even though Starbucks' target markets have broadened greatly since it began as an urban coffee shop serving downtowns and trendy gentrifying neighbourhoods, they are still limited by population size and largely tied to wealthy areas. Starbucks is less interested in servicing poorer neighbourhoods in urban areas, or small town or rural markets.

Independent coffee shops, with their smaller operations and lower capital investments, are not averse to smaller, less affluent markets. They locate in innovative spaces, notably small towns and rural areas that corporate coffee does not wish to, or cannot service. Even in larger markets, the small coffee shops will locate in areas not adopted by their corporate rivals. They locate in less affluent areas, and perhaps most innovatively, locate in the underutilized spaces of the city, such as on small lots, on parts of larger lots, or even in existing parking lots. Since the physical structures of the small shops are insubstantial, they may be easily torn down, or even moved as the market changes, or if the landlord terminates the lease.

It is impossible to determine if either small or large coffee enterprises are being more innovative in their business strategies since they are innovating in different ways and within distinct operating and market conditions. Objective measures are not available to rank the practices uncovered in this research, furthermore disallowing one to say whether the corporate or independent shops are being more innovative. What can be stated is that each must be successful in their innovations since they are able to survive and thrive in a competitive market.

Further Research

Exploring the relationship between the small, independent coffee shops and the large, corporate players offers much insight into the business practices of enterprises at various scales. By adopting innovative strategies, as well as some inherent benefits of their nimble and flexible scale of operations, the small coffee shops are able to compete with their large counterparts. Much ink has been spilled in regards to the corporatization through agglomeration in the retail and service sectors, wherein the small ma and pa enterprises are increasingly pushed out by the corporate chains (for example Mitchell 2000). In the coffee industry we see rather than monopolization by the big players, a relatively peaceful cohabitation occurring, wherein each is competing in the market quite successfully.

In fact, the smallness of the independents may be what makes them competitive in the gourmet coffee market. Their diminutive size allows for two significant benefits over their corporate competitors: capital investments are limited, thus permitting many new start-ups to enter the market, as well as existing operations to innovate when the costs of failure are low. Second, being small means that they are more flexible, responding to market demands much more quickly, thus eliminating the problems that large corporate players must contend with due to the intrinsic lethargy in large bureaucracies.

Stemming from these observations are further avenues of research regarding innovation in the coffee shop industry specifically, and the retail and service industry in general. The arguments made here for how small coffee shops are able to compete with their large corporate rivals should be tested in other geographic regions, and in additional industries. Doing so will determine if the patterns are occurring elsewhere, and evaluate the current state of the retail and service sectors. Furthermore, additional work will assist in predicting future trends, especially to determine if corporatization and homogenization's logical end will be complete domination by corporate chains in most segments.

In terms of coffee specifically, more work is needed as to how consumption patterns have changed, leaving the confines of its origins in trendy city centre locales and entering new spatial markets. The spatial changes are accompanied by demographic changes, wherein gourmet coffee is no longer exclusively consumed by members of the upper class (creative class, bobos). Gourmet coffee is now consumed in various formats of coffee shops and cafés by many new social groups. These emerging consumption patterns illustrate both innovations in marketing and delivery of coffee as well as consumers' cultural shifts.

A survey of the various players in the coffee shop industry would provide rationale for their business strategies, and uncover further innovations in their business practices. Financial information pertaining to capital investments, revenues, and profit would be extremely telling, but it is unlikely that the businesses would offer such valuable information. A survey would, however, reveal the perceptions of the independents and corporate players as to how they view their competitors, and motivations behind strategies adopted. Results will allow us to gain further insight into how the independent coffee shops are able to successfully compete with their corporate rivals, and how both large and small players continue to innovate to win customers.

References

Berry, B.J.L. 1963. *Commercial Structure and Commercial Blight, Retail Patterns and Processes in the City of Chicago.* Chicago: University of Chicago – Department of Geography Research Paper, 85.

Bloom, P.N., and Perry, V.G. 2001. Retailer power and supplier welfare: the case of Wal-Mart. *Journal of Retailing*, 77(3), 379–96.

Brooks, D. 2000. *Bobos in Paradise; The New Upper Class and How They Got There.* New York: Simon and Schuster.

Bryant, S. 2009. *Everything but the Coffee: Learning About America from Starbucks*. Berkeley: University of California Press.

Conzen, M.P., and Conzen, K.N. 1979. Geographical structure of nineteenth-century urban retailing: Milwaukee 1836–1890. *Journal of Historical Geography*, 5(1), 45–66.

Cow Girls Espresso. 2013. [Online] Available at: http://www.cowgirlsespresso.com/ [accessed: 5 January 2013].

Domosh, M. 1990. Shaping the commercial city: retail districts in nineteenth-century New York and Boston. *Annals of the Association of American Geographers,* 80(2), 268–84.

Efrati, A., and Gasparro, A. 2012. Starbucks Invests in Square. *Wall Street Journal* [Online 8 August] Available at: http://online.wsj.com/article/SB1000087239639 0444423704577575803898185594.html [accessed: 3 January 2013].

Ferdows, K., Lewis, M., Machuca, J., and Laurent, R. 2003. Zara. *Supply Chain Forum*, 4(2), 62–67.

Fishman, C. 2006. *The Wal-Mart Effect: How the World's Most Powerful Company Really Works – And How It's Transforming America*. New York: Penguin.

Florida, R. 2002. *The Rise of the Creative Class and How it's Transforming Work, Leisure, Community, and Everyday Life.* New York: Basic Books.

Frenkel, S. 1995. A pound of Kenya, please or a single short skinny mocha. *Pacifica: The Association of Pacific Coast Geographers*, Fall 1995, 12–15.

Gumprecht, B. 2008. *The American College Town.* Amherst: University of Massachusetts Press.

Hernandez, T. 2007. Enhancing retail location decision support: the development and application of geovisualization. *Journal of Retailing and Consumer Services*, 14(4), 249–59.

Hilletofth, P., and Hilmola, O.P. 2008. Supply chain management in fashion and textile industry. *International Journal of Services Sciences*, 1(2), 127–47.

Jones, K., and Simmon, J. 1990. *The Retail Environment*. New York: Routledge.

Lees, L., Slater, T., and Wyly, E. 2008. *Gentrification*. London: Routledge.

Longstreth, R. 1997. *City Center to Regional Mall: Architecture, the Automobile, and Retailing in Los Angeles, 1920–1950*. Cambridge: MIT Press.

Mitchell, S. 2000. *The Home Town Advantage: How to Defend Your Main Street Against Chain Stores – and Why It Matters*. Institute for Local Self Reliance,

Noren, D.L. 1990. The economics of the golden arches: a case study of the McDonald's System. *The American Economist*, 34(2), 60–64.

Novak, M., and Gilliland, J. 2011. Trading places: a historical geography of retailing in London, Canada. *Social Science History*, 35(4), 543–70.

Oldenburg, R. 1999. *The Great Good Place: Cafés, Coffee Shops, Bookstores, Bar, Hair Salons, and Other Hangouts at the Heart of a Community*. New York: Marlowe.

Pendergrast, M. 2010. *Uncommon Grounds: The History of Coffee and How it Transformed Our World.* New York: Basic Books.

Roseberry, W. 1996. The rise of the yuppie coffees and the reimagination of class in the United States. *American Anthropologist*, 98(4), 762–75.

Schultz, H. 1997. *Pour Your Heart Into It: How Starbucks Built a Company One Cup at a Time*. New York: Hyperion.

Smith, N., and Williams, P.R. 1986. *Gentrification and the City*. Boston: Allen and Unwin.

Thompson, C.J., and Arsel, Z. 2004. The Starbucks brandscape and consumers' (anticorporate) experiences of glocalization. *Journal of Consumer Research*, 31, 631–42.

United States Bureau of the Census Bureau. 2010. *American Community Survey, Table CP04, 2010.*

Wurgaft, B.A. 2003. Starbucks and rootless cosmopolitanism. *Gastronomica: The Journal of Food and Culture,* 3(4), 71–5.

Chapter 10
The Role of Creativity and Innovativeness of Firm Managers in Socio-economic Development: The Case of the Poznań Metropolitan Region

Michał Męczyński

Introduction

As Pratt (2008) points out, not only innovativeness but also creativity plays an important role in the knowledge based economy. Much attention has been paid recently to both of these terms in discussions among scholars and policy makers. Many publications in socio-economic geography and economics focus on innovativeness, its dynamics and spatial differences. However, thanks to Florida's (2002, 2005) research, the interest in creativity and its role in the modern economy have increased, too.

In his opinion creativity, i.e. creation, is one of the most important sources of socio-economic development, including establishing new companies and creating new workplaces. A new socio-economic stratification occurs as a result of this development where the so-called creative class occupies an increasingly important position. Its members get involved in work with an aim to create new forms, which is a unique feature of this class, and a significant part of the added value of this kind of activity is its non-material character. Representatives of this class include businessmen, directors and managers of companies. Both creativity and innovativeness play an important part in the activity carried out by them.

Schumpeter (1912) at the beginning of the twentieth century pointed out the importance of human activity for economic development. He emphasised a significant role of individual business people who are promoters of 'creative destruction' and developmental processes which they are responsible for. The entrepreneurs are not only capital owners but also have specific characteristics, most important of which is entrepreneurship. The core of their activity is 'creative destruction' of the economic balance. Thus, economic development is created. From the perspective of the examined issue of creativity, an important characteristic of entrepreneurs is a practical skill to take advantage of the continuous stream of innovations which are at their disposal. Their creative thinking abilities allow them to implement innovative solutions in the economy. However, implementing these solutions entails a big risk of failure in case of a too small number of consumers on the market and the loss of capital invested in research and development (O'Sullivan et al. 2006).

Nevertheless, these innovations are developed, not only because of the advanced skills of the creative sector representatives. Also the environment in which these individuals live and work has a significant impact on the creative process (Landry 2006). So far it has been assumed that the environment which facilitates the development of creativity and innovativeness should be characterised by a favourable business climate, which basically means low taxes and good material infrastructure. However, Florida (2002) reverses this link and claims that a 'good social climate' is what attracts creative people, entrepreneurs, managers and innovative enterprises which come together with them. Representatives of the creative sector do not migrate to specific places only because of a high rate of economic development. The correlation is quite opposite – a high rate of economic development is the result of the presence of large clusters of the creative class in particular places. Florida claims that paradoxically it is not people who migrate to find a job but jobs migrate to find people (Stryjakiewicz and Męczyński 2010, Stryjakiewicz and Stachowiak 2010).

One of the most important factors that determine the character and attractiveness of the place, which is an environment conducive to creativity and innovativeness, is the political and economic system of a country. The system affects the relations between creative individuals and the underlying principles of the economy, among others. Paths of urban development indicate that the past political or economic system of a given state played a crucial role in its current competitiveness (Stryjakiewicz, Burdack and Egedy 2010). Poznań is one of the cities whose character has been affected by the Second World War and almost 50-year rule of the communist regime. The five years of the German occupation (1939–1945) and 50 years of communist rule saw a deteriorating socio-economic situation of the Poznań metropolitan region (PMR). The development of the region along with other metropolitan regions in East Central Europe between 1945 and 1989 strongly relied upon all-embracing, generic social and economic policies formulated by the central government. Prior to 1989, the state's socialist political structure and the centralised management of society practically eliminated the possibility of the development of individuals' creativity. The state's socialist/communist system downplayed by its essence the importance of competitiveness (Stryjakiewicz et al. 2013). However, Poznań did not become a heavily industrialised city and developed commercial functions. Those are both seen as positive legacies in the present phase of economic and social development and leave the city without the difficult task of dealing with structural economic change that affects some other cities. One response to the economic failures of this period was the development of the Independent Self-governed Trade Union Solidarity in the early 1980s. This movement contributed to releasing the creative potential of both the country and the city and led to the political changes of 1989. The new political and economic situation allowed the rebirth of entrepreneurship and organisational skills among the residents of Poznań and restored the city's traditional openness associated with its commercial and communication functions (Stryjakiewicz et al. 2007).

The discontinuity of the city's development path and its inability to keep up with global processes under the communist system has serious implications for the development of individuals' innovativeness and creativity. The development path has left the city with a diversified economy with a dominating service sector, efficient institutions and well-managed city finance, a tradition of entrepreneurship and high work standards, a high quality of life, attractive housing and a rich cultural milieu (especially music and dance; Stryjakiewicz et al. 2013, Stryjakiewicz et al. 2007,). Nowadays, Poznań is one of the most dynamic cities in Poland. With its 568,000 inhabitants it is the fifth largest city in the country after Warsaw, Cracow, Łódź and Wrocław (the Poznań Metropolitan Region has a population of 856,000). The ethnic structure of Poznań is highly homogeneous with only 0.2 per cent of non-Polish nationals (Stryjakiewicz et al. 2007). The geographical position of the city, located in the west-central part of the country and equidistant between Warsaw and the German capital Berlin, puts it under a strong competitive pressure from both capitals (Brown and Męczyński 2009: 233). In 2005 the number of employees in the creative sector (including creative industries and knowledge-intensive industries) in the PMR was 94,260 (17.5 per cent of total employment). In the same year the number of creative entities was 23,985 (17.6 per cent of the total number of all entities; Stryjakiewicz and Stachowiak 2010).

In this publication the author attempts to determine the role played by creativity and innovativeness of the creative sector managers in the socio-economic development of the Poznań metropolitan region. The characteristic features of the environment conductive to the activity of creative individuals were taken into consideration when analysing the complexity and dynamics of the changes which occur in the city struggling with the burden of the post-world war development path. Therefore, the author seeks to answer four research questions, which are as follows:

- What is the relation between creativity and innovativeness and what role is played by the environment in their development?
- How significant role do some of the managers play in the development of creativity and innovativeness in firms in the PMR?
- What are the common and unique features of an environment (milieu) facilitating creativity and innovativeness?
- What is the influence of firm managers on socio-economic development of the PMR?

The chapter consists of five sections. The first is of an introductory character. It outlines the research issue and the aim of the article. The second section discusses the relations between innovativeness and creativity, taking into consideration the activity of the creative sector representatives and the role played by the environment for their development. The next part is devoted to the research methodology. Then, the fourth section presents the results of a research conducted among the creative sector representatives concerning their role in socio-economic development of the Poznań metropolitan region. The last section contains a summary of the findings.

Relations between Creativity and Innovativeness and the Role the Environment Plays in Their Development

The issues of both creativity and innovativeness are extensively discussed in the literature of various scientific disciplines. When addressing the creativity and innovativeness of managers, it is important to establish the relation between these two notions. The first issue was pondered at length in psychology (Table 10.1). It is especially important to emphasise the great input of Guilford (1950). He concluded that creativity was the generation of multiple answers to an open-ended problem on the basis of divergent thinking, characterised by the criteria of fluency (ability to produce a large number of ideas), flexibility (capacity to consider a number of different approaches simultaneously) and originality (tendency to produce ideas and solutions that are untypical, unusual, unique). In turn, from the perspective of human geography, Törnquist (2004) firmly indicates the significance of the environment, the so-called creative milieu, in defining creativity. He connects this notion with a process of generating ideas that takes place in the mind of an individual, both in isolation and in interaction with the surroundings.

Researchers from other fields see creativity as a process (Stein 1953, Mooney 1963, Törnquist 2004), or a result of a creative activity which occurs as a material or non-material product (Stein 1953, Mooney 1963). Finally, Florida (2002) indicates individuals – creative class representatives, who conduct creative activity.

The issue of innovativeness was also addressed by numerous researchers. Schumpeter (1912) articulated the most popular and general definition of innovativeness (Table 10.2). It has laid a basis for defining operational innovation for the needs of contemporary statistics (OECD 2005). In turn, Amabile (1996) stresses the development of innovativeness as a process.

A feature common to both types of activity is creating a product. In the case of creative activity it is of a material or non-material character, whereas in the case of innovativeness it is mostly a market-oriented material product. The analysis of both notions presented above leads to the conclusion that creativity is a slightly broader notion and is primary with respect to innovativeness. Simultaneously, a creative individual such as a manager or entrepreneur plays a crucial role in the transition between these two stages (Figure 10.1).

Table 10.1 Definitions of creativity – selected fields of science

Field of science	Author of definition	Chief aspect of definition
Psychology	Guilford (1950)	Fluency, flexibility, originality
	Stein (1953)	Relative character of novelty and value of human product
	Mooney (1963)	A process, surroundings and the work's author
	Nęcka (2005)	Conjunction of novelty, value and behaviour of a creative person
Sociology	Florida (2002)	Creative individuals responsible for the development of metropolitan centres
Human geography	Törnquist (2004)	Creative surroundings (creative milieu)
	Landry (2006)	Significance of creative social institutions in economic development

Source: author's compilation.

Table 10.2 Definitions of innovativeness – selected fields of science

Concept of creativity		
Field of science	**Author of definition**	**Chief aspect of definition**
Economics	Schumpeter (1912)	A new good, production method, market, supplies, organisation
	OECD (2005)	A definition introduced for the purpose of international statistics embraces innovations in a product, process or organisation
Management	Amabile (1996)	A process, a product, a market
Human geography	Asheim and Gertler (2006)	A significant role of tacit knowledge connected with a concrete place and persons

Source: author's compilation.

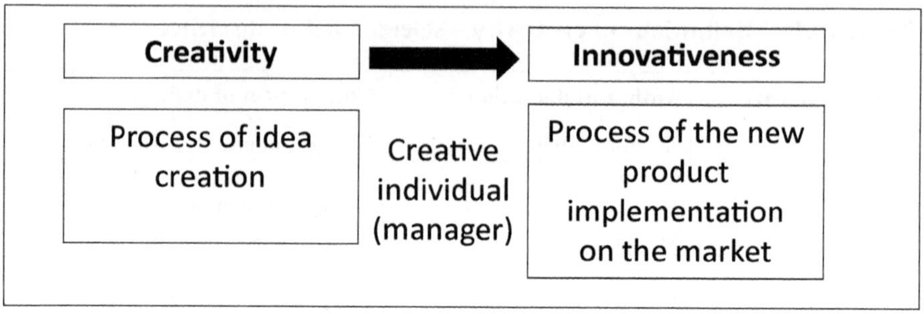

Figure 10.1 Role of creative and innovative firm managers in the process of idea creation and its implementation in the market
Source: author's compilation based on Amabile (1996) and Ekvall (1997).

Drucker (1985: 27) states, drawing on Schumpeter's theory: 'Dynamic disequilibrium brought on by the innovating entrepreneur, rather than equilibrium and optimisation, is the norm of a healthy economy and the central reality of economic theory and practice'. Both the invention of ideas and their implementation on the market are the domain of creative individuals capable of getting people to accept a novelty using a network of social contacts and bringing about its sale on the market.

At the same time, these two issues are related to other types of activity. As far as creativity is concerned, it might be artistic creation, whereas in the case of innovativeness it might be an activity which involves intensive use of knowledge.

However, creative individuals do not act in isolation from their surrounding environment. The factors influencing the decision of choosing a location of activity by creative individuals, increasingly taken into consideration in research, include the quality of the environment connected with the place of living or working. It is connected with moulding the so-called creative milieu which is characterised by good interpersonal relations between employees in the workplace or between neighbours within the same living community (Törnquist 2004). Representatives of the creative sectors can get deeply ingrained in the region on condition that they have an appropriate amount of time necessary to establish good networking relations both with co-workers and neighbours. From the perspective of the economic development of the Poznań metropolitan region, it is crucially important to establish networks of businesses in which representatives of the creative sector may function. Business networking issues were addressed by Stryjakiewicz (2004), who assumes, on the basis of literature studies, that the notion of a network refers to 'a stable group of relations between actors – participants of the economic game'. According to this trend, the relations which establish and identify networks occur between actors – participants of the economic game and are fostered by such factors as

a) competition, b) cooperation, and c) control. Lately network relations have become particularly important in companies where flexible networks replace top-down structures. Also the traditional department structure in companies is being replaced by temporarily active project teams which gather specialists from the creative sector.

The creative environment in the workplace is moulded by individuals who have different personalities. They create an organisation consisting of people who cooperate with one another in an orderly and coordinated manner to achieve a set of goals (Griffin 1996).

The results of numerous studies indicate that understanding mechanisms underlying the changes introduced in an organisation of companies is not possible without understanding the rules underlying small social groups, in company departments or research units (Törnqvist 2004). The character of work conducted and its results might be different. While managers connected with creative activities contribute to the development of creativity, people connected with activities which involve intensive use of knowledge generate innovations. However, both in the former and the latter case, good working conditions in the workplace (the ability to network and exchange information, among others) will play a crucial role in increasing efficiency.

The author of the creative milieu concept – Törnqvist, analyses the issue of creativity in his work entitled 'Creativity and the renewal of regional life' (1983) where he refers it to the scale of the city or the region. He notes that one can talk about benefits connected with the following aspects: 1) location (proximity of wildlife, among others), 2) an institutional environment fostering the development of entrepreneurship (a clear tax system, among others), and 3) a friendly climate for development of cooperation between employees (facility in establishing cooperation, among others). Such an environment is conducive to implementing creative solutions. It leads to increasing competitiveness between businesses and consequently cities and regions where they are located (Malecki 1987).

Research Methodology

The information about the role of creative and innovative managers in the socio-economic development of the Poznań metropolitan region was obtained from surveys (conducted among employees of the creative sector) and individual in-depth interviews (involving managers and foreign immigrants connected with the creative sector). The research was carried out within the framework of two international projects, ACRE (Accommodating Creative Knowledge. Competitiveness of European Metropolitan Regions within the Enlarged Union) and CREA.RE Creative regions.

The former project, ACRE, was carried out in the period of 2006–2010 and its aim was to assess the impact of the emerging 'creative class' and the rise of the 'creative industries' on the competitiveness of EU metropolitan regions The project was funded within the Sixth Framework Programme of the EU and embraced scholars from 13 metropolitan regions in Europe (Amsterdam, Barcelona, Birmingham, Budapest, Dublin, Helsinki, Leipzig, Milan, Munich, Poznań, Riga, Sofia, and Toulouse). The team from Adam Mickiewicz University in Poznań was supervised by Professor Tadeusz Stryjakiewicz (ACRE 2013) .

An important part of the project was to evaluate the opinions of the members of the three target groups: (1) employees, including graduates, 2) managers, and 3) trans-national migrants, about the role of 'soft' location factors in business location strategies as compared with other location factors, the role of the historical development path of an urban economy, the urban social, cultural and physical structure, and significant differences regarding the opinions of three target groups.

Altogether 200 employees of the creative sector (100 representatives of creative activities and 100 representatives of knowledge-intensive businesses) took part in the survey studies conducted under this project in the Poznań metropolitan region in July and August 2007. The latter group also included 50 graduates of selected faculties of universities and artistic schools of the tertiary education level. In addition, this article uses research results based on individual in-depth interviews, also conducted within the ACRE framework. A total of 50 persons participated in them, including 20 managers of the creative sector, 25 international migrants, and 5 experts dealing with the issues of migration of creative individuals with high qualifications. The interviews were conducted in March 2008 and 2009. A detailed description of the sample group selection for the survey studies and in-depth interviews within the ACRE framework is presented in Stryjakiewicz and Męczyński (2010).

The second project – CREA.RE Creative Regions – made it possible to update and extend the knowledge of the creative sector in the PMR. The aim of the project was to exchange experiences and make common efforts for a better, stronger and sustainable embeddedness of cultural and creative industries in regional policies. The CREA.RE partnership comprised 12 local and regional public authorities from 10 EU countries (Austria, Finland, Germany, Italy, Poland, Romania, Slovenia, Sweden, Spain, and Belgium). The implementation of the project involved conducting 10 in-depth interviews with representatives of the creative sector in January 2012. The aim of the interviews was to collect information about problems that creative firms' managers have in running their businesses in the Poznań metropolitan region. It was also possible to collect their opinions about possible solutions.

Selected Representatives of the Creative Sector and Their Role in the Poznań Metropolitan Region (PMR)

In 2009 managers in the creative sector in the institutions and companies of the Poznań metropolitan region accounted for 5 per cent (4,713 people) of the total number of employees in this sector (Stryjakiewicz, Stachowiak 2010).

Survey studies carried out under the ACRE project made it possible to indicate the most creative managers in the Poznań metropolitan region. The respondents mostly associated creativity with artistic and cultural activity. The group of 11 people indicated by the respondents included mainly managers who represent the subsector of artistic activity. Only four of them were connected with the sector involving knowledge-intensive businesses. However, also their activity was frequently conducted on the border between business and art.

One of these people, and the one mentioned most often, was Grażyna Kulczyk, who together with Jan Kulczyk was considered the most creative manager. Ms Kulczyk has been the founder and only shareholder of the Fortis company since 1998. It was created to modernise and restructure an old brewery into a modern Stary Browar ('Old Brewery') Trade, Business and Art Centre. At present, the company is responsible for managing and planning further development of the Old Brewery. The quality and class of the Old Brewery has been confirmed by international awards (award of the International Council of Shopping Centers for the Best Shopping Center of the World and the International Council of Shopping Centers for the Best Shopping Center in Europe, both in the category of medium-sized shopping centres; Stryjakiewicz et al. 2007: 61). Ms. Kulczyk has been the animator and patron of many artistic and cultural events in the city, like sculpture and painting exhibitions, theatrical and opera performances, or musical concerts. She collects art. Part of her collection, which has been amassed for over 30 years, is exhibited in the Old Brewery modern art gallery. The activities conducted by Ms. Kulczyk, which have influence on the economic development of the PMR, involved engaging 27 firms at the stage of project development and its implementation. It led to creating 400 new work places. To begin with, 1,300 persons found jobs in the part of the building opened first – the Old Brewery Atrium. After opening the second part in 2007 – the Old Brewery Passage, the total number of employees reached 2,500, while in 2012 employment increased to 4,000. In running the Old Brewery, Ms. Kulczyk combines business activity with animation and patronage of artistic and cultural events in the city (such as sculpture and painting exhibitions or opera and theatrical performances, or musical concerts; Stary Browar 2012). Part of Grażyna Kulczyk's collection belongs to the exhibition of the modern art gallery in the Old Brewery.

Jan Kulczyk, one of the richest people in Poland (Ranking 100 najbogatszych ..., 2013), also belongs to people who contribute to increasing the competitiveness of the Poznań metropolitan region. He is the owner of Kulczyk Investments, a parent company registered in 2007 in Luxembourg. He manages the companies in the holding which conduct business internationally in the most attractive sectors of the economy, such as the power industry, mineral resources, infrastructure, and real estate. The Kulczyk Holding, a subsidiary of Kulczyk Investments, embraces 22 companies which are active in Poland (Kulczyk investments 2012). It participates in charities through the activity of the Kulczyk Family Foundation, which funds scholarships for students and young scientists from Poznań.

The group listed by the survey respondents included entrepreneurs running modern knowledge-intensive companies, such as those connected with the production and designing of buses (Solaris Bus & Coach, until 2001 it was Neoplan Polska LLC) which are purchased both in Poland and abroad, for example in Germany, Estonia and the Emirates; or the designing and production of furniture (the Vox Group of Companies). The first group of the selected representatives of the creative sector combined business activity with supporting culture, whereas the other group was connected mostly with creative activities which were their main source of living.

The respondents mentioned the following people as the most creative managers of artistic activity: the Director of the Polish Theatre of Dance, Poznań Ballet – a graduate of L'Academie de la Danse in Paris (Culture.pl 2012), the former director of the International Malta Theatre Festival who was in charge of the street theatre festival which contributed to the growth of the importance of Poznań as a cultural centre in the 1990s, an architect responsible for creating the most characteristic buildings in the city, and the foreigner who established the Media Rodzina publishing house, which has published in Polish such books as *Chronicles of Narnia* by Lewis and *Harry Potter* by Rowling. The group of recognisable persons included not only those who were good managers, but also social activists who run different social events conducive to the process of revitalising the city of Poznań, and clergymen connected with the organisation of national youth prayer meetings.

Activities Undertaken by Creative Managers Facilitating the Development of their Institution or Company

The previous studies dedicated, among others, to the identification of factors attracting creative individuals to Poznań and its metropolitan region have indicated the possibility of finding employment as one of the major decisive elements (Stryjakiewicz and Męczyński 2010, Stryjakiewicz and Stachowiak 2010). This perspective was a direct result of, among others, a good organisation structure of businesses connected with the creative sector; they generated new jobs and in this way contributed to a steady increase in the level of social and economic development of the city.

The participants of the study belonged to a group of people managing public institutions of culture important for Poznań and its metropolitan region. The institutions were strongly connected with the history of the city and at the same time enjoyed a worldwide renown thanks to their participation in numerous cultural events.

> The seat of the 'Zamek' Cultural Centre is a 1910 imperial castle built for William II. It has been used as a cultural venue since 1962, when its function was changed from the seat of the City National Council to the Palace of Culture. This institution for the dissemination of culture is the biggest in Poznań

> and one of the biggest in Poland. Each year it provides opportunities for the organisation of around 700 events: exhibitions, concerts, movie screenings, theatre performances, meetings, and education workshops. (Representative of the Cultural Centre)

The creative activity of managers from the Poznań metropolitan region who participated in the study, particularly those connected with film animation, was honoured by awards. Participation of these institutions in international events where they received those awards brought substantial acclaim to both, them and the city and the region.

> The company produces animated films [...]. It has received over 180 national and international awards, for example the 2011 Awards of the International Jury for the Best Polish Film at the International Festival of Films for Young Viewers!' in Poznań, the Silver Seal of the City of Poznań in 2011 for its contribution to culture and film art. (Representative of the TV Studio of Animation)

Although managers participating in the interviews who represented the public sector were recognised internationally, they had to struggle with financial shortages in Poland. Such a situation occurs in particular in the case of financing cultural and creative activities. Therefore, they undertook creative activities in order to raise funds. As a consequence, some of them tried to obtain external resources from domestic sources (for example private donors and ministries) and international ones (for example the European Union's structural funds). External resources were obtained by managers representing both public and private entities. Still, the actions taken by the latter group of entities seemed more efficient. In the public sector, resources were obtained for projects meant to subsidise tangible infrastructure. However, managers also sought funds for projects connected with the coordination of actions of and consulting for companies, including those connected with the creative sector.

> The biggest project implemented by the Centre of Culture is the 'Reconstruction of the complex of the historical Grand Chamber of the Centre for an effective use of cultural heritage'. The value of the project is 12 million Euro (including part-financing by the European Union). (Representative of the Cultural Centre)

The creativity develops in Poznań also thanks to private investors. A case in point is the centre of design and business creativity, the Concordia Design, an incubator for start-ups in the creative sector, including advertising agencies and architectural studios. This is a place where young artists can develop their talents. The managers of the Centre assist in promotional actions and offer a chance to open up own offices or studios at preferential rates. It holds workshops during which young artists come in contact with renowned representatives of the art scene, in particular design. The operation of the Centre is in large measure part-financed from the European Union's structural funds.

> The Centre is located in the heart of Poznań, in a regenerated historical building of a former printing house. It cooperates closely with the School of Form affiliated with the School of Humanities and Journalism and the School of Social Psychology. The principal operation of the Centre can be divided into three types of actions: 1) education, workshops, trainings and conferences concerning, among others, the management of the product development process for individual commissions and design in public space, 2) consulting, or the provision of advisory services concerning all aspects of designing, product development process, design management, analysis of trends in the sector, and consultations concerning market research, and 3) promotion, or the organisation of design competitions and the promotion of Polish and international design'. (Representative of the Concordia Design)

Building a network of links with the surroundings is a long-term process based on both formal and informal contacts which often exist even before a company is started. Spatial proximity in cities favours development and the creation of various networks embracing those managers as well as non-market interactions based on direct customer relations (Stolaric, Florida 2006). The quality of these relations to a large extent determines the possibilities of development of the creative potential of an individual, who often runs a business or heads an institution. Such personal skills in establishing good contacts and creating proper relations with the environment frequently determine the operation of a company or an institution (Bathelt, Malmberg and Maskell 2004).

The effects of globalisation and extensive chances of information transfer by electronic media are taken as an advantage by representatives of companies and institutions related to the private sector, for example in Concordia Design. They use all kinds of social networking services (for instance Facebook) and websites to disseminate information about their activities, and this establishes an online community centred around these companies.

> We make use of various channels to reach out to potential clients. The most important of these channels are as follows: a Facebook profile with separate sub-pages related to the major events co-organised by 'Zamek', radio advertisements and (less often) TV commercials, posters and pamphlets (distributed in clubs, at universities, in institutions of culture, outdoor advertising, for example on the western façade of the Centre's tower), contact with the media, dedicated invitations, and organisation of press conferences. There are plans to launch a new website, which will be connected with a call for proposals for the visual identification of the Centre, upon the completion of the renovation of the Grand Chamber. (Representative of the Cultural Centre)

Managers representing the creative sector in Poznań spare no effort to make use of ICT to create good relations with the external environment. They use such technologies to disseminate information about cultural events taking place in Poznań. Some managers conduct promotional activities as their main line of operation, like those from the Office for the Promotion of the City, who organised a lot of actions presenting the economic potential of Poznań-based brands. This contributed to the creation of a new image of the city within the framework of the project known as the Consortium of the Brand of Poznań.

Moulding a Creative Environment in the Workplace

Florida (2002) claims that most employees of the creative sector can be considered an example of 'role model entrepreneurs'. One of the signs of independence of entrepreneurs who belong to the creative sector is the mode of conducted work. In many cases it is not regulated by fixed working hours. Simultaneously, work in the creative sector involves a great deal of freedom as regards internal work organisation and implementation of one's own ideas (Gill and Pratt 2008). These aspects of work were often perceived as beneficial by the employees who participated in the study.

The conducted surveys and interviews made it possible to determine the influence of activities carried out by managers from the creative sector on the conditions of work, both theirs and their employees'. The changes they implemented in their organisations aimed at increasing the quality of employment conditions and establishing a creative environment in the workplace, which was supposed to contribute to the economic development of the region.

As a result of the survey studies it has been established that temporarily employed workers accounted for 43 per cent of employment in the Poznań metropolitan region (contracts for a specified period, for example for the implementation of a project, agreements for performance of a specified task, or commission contracts). More than one-third of the respondents were self-employed (31.6 per cent). Simultaneously, men were more inclined to accept this form of employment (30.7 per cent) than women (20.9 per cent). Women preferred more stable employment because only slightly above 30 per cent of them were employed temporarily (Brown, Nadler and Męczyński 2010). It created more favourable conditions for starting a family, having and raising children, because almost 70 per cent of women working in the creative sector could take advantage of welfare benefits offered by the state thanks to being employed for an unspecified period of time. The possibility to obtain various benefits was highly appreciated not only by women but also by most (55.3 per cent) of those surveyed. However, for the participants the possibility to control their work was the biggest advantage (77.6 per cent of those surveyed).

Table 10.3 Evaluation of the creativity of managers in the Poznań metropolitan region

Aspect of creativity	Strength of aspect*	Managers' opinions
An aim of the company is a challenge for the manager	++	"It has been observed recently that TV shapes aesthetic taste […]. Hence there is a need for series that would stand out, would be different. Something that would not be a replica of what has been previously created. Something that would not imitate the same aesthetics and vividness. […]. And I think this is what determines our strength. The power of creativity".
Freedom of actions	++	"We don't have films […] which imitate the same idea. […]. This possibility of artistic expression is highly motivating for young people".
Co-workers' support in implementing new ideas	+	"It isn't a big company, it isn't a corporation, it is still a family firm. I try to prevent the rat race and put a stop to any person trying to establish that atmosphere. Trust is important, especially regarding the people who shape the firm's activity. They should cooperate and understand one another well. I try to resolve all tensions, if possible in a friendly manner, sometimes in a top-down fashion". "Naturally, much depends on who you co-operate with, but here Poznań differs a bit, I think, from other cities in that people approach business very seriously". "Partner integrity, that's what's the most important here […]".
No fear of coming up with a new idea	–	"[Young graduates] are so closed to the market that they are unable to take up a freelance job in a professional studio with a professional band. As if it was a shame for artists to work for money".
Possibility of discussion in a friendly atmosphere	++	"We also discuss some tax questions, growth questions of the city itself, new facilities being built in the city and possibilities of conducting our activity in those facilities".
Tolerance	+	"Over the last three or four years we can observe firms from outside. From time to time we have commissions from, say, Płock, or from Silesia, from Wrocław, but they turn up less and less frequently. Right now the dominant market is that of Poznań, and more precisely, Wielkopolska".

Note: * ++ very strong, + – strong, – minor.
Source: author's compilation based on Nęcka (2005) and ACRE project interviews (50 respondents).

Despite the high level of employment insecurity and risk of losing work, a big number of highly qualified, mostly young, people look for a job in the creative sector, particularly related to artistic activity. They are attracted by an image of a dynamic and enterprising person who has a great deal of autonomy in the activities undertaken, which is associated with work in this sector. At the same time, work in the creative sector gives many possibilities of self-fulfilment and employing one's skills. The character of work performed is prestigious because, according to many people, only a few are talented enough and have unique qualifications to belong to the group of the so-called 'chosen ones' (McRobbie 1998).

The work of a manager from the creative sector is also connected with emotions which he or she is exposed to when doing the job. Kocowski (1991) lists emotions which stem from the atmosphere in the workplace and contribute to creating new ideas: curiosity, exhilaration, interpersonal understanding, and other positive emotions fostering activity.

The results of the interviews conducted in companies and institutions in the Poznań metropolitan region lead to the conclusion that the atmosphere created by co-workers in the workplace evoked positive emotions. They contributed to a strong commitment of the managers to achieve the goals they faced (Table 10.3). They treated them as challenges which they had to take up and devote their whole creative potential to them. A crucial development factor was the feeling of freedom in undertaking new activities. However, some of these emotions connected with the workplace had a negative influence on the people in charge of business entities and made them fear presenting their own plans or projects and being rejected. This situation took place despite the fact that they described the atmosphere in the workplace as free enough to encourage discussion about new activities.

Summary

This article addresses the issue of the role played by creativity and innovativeness of managers in the creative sector in the economic development of the Poznań metropolitan region. The first part discusses the relation between creativity, innovativeness and the environment, known as a creative environment, or milieu.

The conducted analysis of the literature leads to the conclusion that the outcome of creative activity consists mostly of products which are of a material and a non-material character, whereas the effects of innovative activity are usually of a material kind. Both the former and the latter contribute to the development of companies and regions. Simultaneously, it has been observed that a representative of the creative sector, such as a manager or entrepreneur, plays a significant role in the transition from creativity to innovativeness. Managers involved in the creative sector become responsible for implementing creative and innovative solutions in the organisation of their companies and institutions intended to increase their competitive advantage. Thereby, they contribute to the development of the region. At the same time these solutions must be adopted gradually and not in an abrupt way. Tapping into the creative potential of the managers who run companies and institutions will be necessary to implement the changes responsibly and predictably.

The studies conducted among the managers and employees of the creative sector in the Poznań metropolitan region within the framework of the ACRE and CREA.RE projects indicate advancing development of both the companies and the region. The environment connected with the place of work of a manager plays a stimulating role in finding creative and innovative solutions.

Problems which restrict the creativity of individuals can be classified into two categories. The first includes issues related to integration and fostering relations with partners from the creative sector in the area of Poznań and its metropolitan region. The other category focuses on the city's policy supporting the creative sector, particularly its artistic activity.

As regards the former issue, it should be observed that one of the features of contemporary socio-economic space is the complexity of relations occurring between the participants of the economic game (such as companies, institutions and organisations). Establishing networks of cooperation based on relations developed between a business entity and its surroundings is a long-term process relying on formal and informal contacts between individuals. Frequently such informal connections, for example family bonds or friendships, become an important factor which influences setting up one's own business.

The spatial closeness of the participants of a network facilitates development of formal and informal contacts between them. As Stolaric and Florida (2006) point out, dense urban space leads to closeness of individuals and makes direct contact possible. These contacts are conducive to creating network structures based on business cooperation. On the other hand, the quality and durability of relations stem from the creative potential of individuals. The developed networks of business contacts determine, to a large extent, the success of an enterprise (Bathelt, Malmberg and Maskell 2004).

Currently, the Internet provides great opportunities to integrate people involved in the creative sector. As far as using this technology is concerned, the managers interviewed demonstrated an interest in the initiative aiming at creating an Internet platform which would serve to integrate various entities involved in the sector in question. The platform could facilitate information exchange between representatives of the sector as regards the enterprises they run and possibilities of cooperation.

Integrating activities of representatives of companies and institutions from the creative sector are also supposed to aid in solving another set of problems, namely those connected with the city's policy of support to the creative sector, a policy seeking to alter the vision of the creative sector and based on collaborative efforts of the creative sector and the local authorities. Poland has no cohesive guidelines for the creative sector that would fit into a holistic strategy of development of the country, the region and the city. There are also no solutions, such as strategic plans, that would facilitate making creative spaces (connected with small architecture and friendly public space) in which representatives of the creative sector could meet and work. Unfortunately, more friendly institutional conditions for the development of creativity, a higher quality of the environment, and a more attractive job offer, mostly in Warsaw, the capital of Poland, lead to a situation where creative young people ('talents') flee from Poznań.

As far as the problem of the drain of young and active people from the city is concerned, the managers of companies and institutions of the creative sector emphasised that the policy of the local authorities should be altered and designed to support activities connected with this dynamically developing sector in particular. It should entail an increase in financial expenditure as well as institutional and organisational improvements. Moreover, it was proposed that Poznań should adopt many solutions already introduced in other cities, not only abroad but also in Poland (for example implementing a policy that would regulate rents on premises that belong to the city and are not leased by creative-sector entities). Adopting solutions which would make it possible to let out premises for this type of activity would help to create working spaces for artists, among others. A greater concentration resulting from spatial density in the city centre would be conducive to the flow of ideas and implementation of collaborative projects, and this would promote the development of the Poznań metropolitan region.

References

ACRE. 2013. Available at: www.acre.socsci.uva.nl [accessed: 15 January 2013].

Amabile, T. M. 1996. *Creativity in Context.* Boulder, CO: Westview Press, Inc.

Asheim, B. T. and Gertler, M. 2006. The geography of innovation: regional innovation system, in *The Oxford Handbook of Innovation,* edited by J. Fagerberg, et al. Oxford: Oxford University Press, 291–317.

Bathelt, H., Malmberg, A. and Maskell, P. 2004. Clusters and knowledge: local buzz, global pipelines and the process of knowledge creation. *Progress in Human Geography.* 28, 31–56.

Brown, J. and Męczyński, M. 2009. 'Complexcities': Locational choices of creative knowledge workers. *Built Environment,* 35(2), 238–52.

Brown, J., Nadler, R. and Męczyński, M. 2010. Working on the edge? Creative jobs in Birmingham, Leipzig and Poznan, in *Making Competitive Cities,* edited by S. Musterd and A. Murie, Oxford: Wiley-Blackwell, 208–31.

Culture.pl 2012. Available at: www.culture.pl [accessed: 4 December 2013].

Drucker, P. 1985. *Innovation and Entrepreneurship. Practice and Principles.* New York: Harper & Row Publishers.

Ekvall, G. 1997. Innovations in organizations. *European Journal of Work and Organizational Psychology,* 5(1), 105–23.

Florida, R. 2002. *The Rise of the Creative Class and How It's Transforming Work, Leisure, Community and Everyday Life.* New York: Basic Books.

Florida, R. 2005. *Cities and the Creative Class.* New York: Routledge.

Gill, R. and Pratt, A. 2008. In the social factory? *Theory, Culture & Society,* 25, 7–8.

Griffin, R. W. 1996. *Podstawy zarządzania organizacjami*. Warszawa: Wydawnictwo Naukowe PWN.

Guilford, J.P. 1950. Creativity. *American Psychologist*, 5(9), 444–54.

Kocowski, T. 1991. *Szkice z teorii twórczości i motywacji*. Poznań: SAWW.

Kulczyk Investments, 2012. Available at: www.kulczykinvestments.com [accessed: 10 October 2012].

Landry, Ch. 2006. *The Creative City. A Toolkit for Urban Innovators,* London: Earthscan.

Malecki, E.J. 1987. The R&D location decision of the firm and 'creative' regions. *Technovation*, 6, 205–22.

McRobbie, A. 1998. *British Fashion Design; Rag trade or Image Industry?* London: Routledge.

Mooney, R. 1963. A conceptual model for integrating four approaches to the identification of creative talent, in *Scientific Creativity: Its Recognition and Development*, edited by C.W. Taylor and F. Barron. New York: Wiley, 331–40.

Nęcka, E. 2005. *Psychologia twórczości.* Gdańsk: Gdańskie Wydawnictwo Psychologiczne.

OECD, 2005. *Oslo Manual*, Paris: OECD.

O'Sullivan, D., Messina, J. P., Manson, S. M. and Crawford, T. W. 2006. Space, place, and complexity science. *Environment and Planning A*, 38(4), 611–17.

Pratt, A.C. *2008. Creative cities*: the cultural industries and the creative class. *Geografiska Annaler: Series B – Human Geography*, 90(2), 107–17.

Ranking 100 najbogatszych Polaków w 2012 roku, 2013, Available at: www.100najbogatszych.wprost.pl [accessed: 10 December 2012].

Stary Browar – Fundacja 5050, 2012. Available at: www.starybrowar5050.com [accessed: 11 December 2012].

Stolarick, K. and Florida, R. 2006. Creativity, connections and innovation: a study of linkages in the Montréal Region. *Environment and Planning A*, 38(10), 1799–1817.

Stryjakiewicz, T. 2004. Sieci gospodarcze w Polsce w warunkach transformacji systemowej, in *Rozwój regionalny i lokalny w Polsce w latach 1989–2002*, edited by J.J. Parysek, Poznań: Bogucki Wydawnictwo Naukowe,

Stryjakiewicz, T., Burdack, J. and Egedy, T. 2010. Institutional change and new development paths: Budapest, Leipzig, Poznan, Riga and Sofia, in *Making Competitive Cities*, edited by S. Musterd S. and A. Murie. Oxford: Wiley-Blackwell: 93–112.

Stryjakiewicz, T., Kaczmarek, T., Męczyński, M., Parysek, J.J. and Stachowiak, K 2007. *Poznań Faces the Future: Pathways to Creative and Knowledge-based Regions.* Amsterdam: AMIDSt, 126.

Stryjakiewicz, T. and Męczyński, M. 2010. *Atrakcyjność poznańskiego obszaru metropolitalnego dla pracujących w sektorze kreatywnym.* Poznań: Bogucki Wydawnictwo Naukowe.

Stryjakiewicz, T. and Stachowiak, K. 2010. *Uwarunkowania, poziom i dynamika rozwoju sektora kreatywnego w poznańskim obszarze metropolitalnym.* Poznań: Bogucki Wydawnictwo Naukowe.

Stryjakiewicz, T., Gritsai, O., Dainov, E. and Egedy, T. 2013. Addressing the legacy of post-socialist cities in East Central Europe, in *Place-making and Policies for Competitive Cities*, edited by S. Musterd and Z. Kovács. Chichester: Wiley-Blackwell, 77–94.

Schumpeter, J. 1912. *Theorie der wirtschaftlichen Entwicklung. Eine Untersuchung uber Unternehmergewinn, Kapital, Kredit, Zins und den Konjunkturzyklus.* Berlin: Auf. Duncker und Humblot.

Stein, M. I. 1953. Creativity and culture. *Journal of Psychology*, 36, 311–22.

Törnqvist, G. 1983. Creativity and the renewal of regional life, in *Creativity and context. Lund Studies in Geography* edited by A. Buttimer, Series B, 50, 91–112.

Törnqvist, G. 2004. Creativity in time and space. *Geografiska Annaler: Series B – Human Geography*, 86, 227–43.

Chapter 11

The Global and the Local in the Polish Biotechnology Industry

Piotr Dawidko

The biotechnology sector is widely considered a "born-global" industry (Geenhuizen and Soetanto 2008). Factors such as close ties with academia, rapid flow of information, and the significant role of the sector's regional megacentres, bringing together leading academic institutions and companies, contribute to the strengthening of this image of the industry (Cooke 2004). At the same time, a variety of local factors also play a very important role in the development of biotechnology, due to one of the basic characteristics of the sector – strong links between science and industry. Local entrepreneurial culture, institutional thickness, and an innovative milieu serve often as factors determining the development of biotechnology enterprises (Lawton-Smith 2004, Lawton-Smith and Ho 2006). The importance of local factors can become even larger, given the sector's strong tendency to form spatial clusters, integrating knowledge centres (universities, research institutes), dedicated biotechnology firms (DBFs), other companies loosely related to the industry, and a wide range of participating organisations (Lawton-Smith 2004).

The interaction between global and local factors tends to be influenced by various determinants depending on the social, political, and economic environment. In the case of Central European countries, specific local conditions exist that affect economic development and innovation – conditions being the result of the region's communist past and the economic and political transformation commenced in 1989 (Malo and Norus 2009, Mroczkowski and Elms 2009).

Taking into account the role of the biotechnology sector as the most knowledge-intensive economic activity and the basis of local innovativeness (Cooke 2004, Gertler and Vinodrai 2009), its development can have a significant impact on Central and Eastern European countries, helping them to catch up economically with Western Europe. Therefore, the role of global and local factors in the functioning of the biotechnology sector in Central and Eastern Europe appears to be an issue worthy of a closer look. The chapter aims to assess this role.

Global and Local Factors Shaping the Biotechnology Sector

The main feature of the biotechnology sector affecting most aspects of its operation is the strong relationship between business and academia (Lawton-Smith 2004). Unlike other high-tech industries, which use the academic sector mainly as a source of trained workers, biotechnology universities and research centres serve as a major source of innovation (Eliasson 2000). The key role of academia in biotechnological innovation is associated with the rapid pace of innovation and progress, which makes it necessary to create highly specialised research teams with access to the tacit knowledge in their field. At the same time, the high uncertainty of success and high costs of maintaining research teams prompt biotechnology entrepreneurs to seek state funds (Coenen, Moodysson and Asheim 2004).

Innovativeness, which is the basis for the functioning and survival of the global biotechnology industry (Mroczkowski, Khilji and Bernstein 2006, Mroczkowski and Elms 2009), is based on the transmission of tacit knowledge occurring only at short distances through personal contacts. Biotechnology companies tend to form geographic clusters around leading research institutions (Gertler 2003, Coenen, Moodysson and Asheim 2004) in order to facilitate this goal. The result is the existence of a global biotechnology cluster system led by several regional research megacentres including the American city of Boston, British city of Oxford, and the German state of Bavaria (Cooke 2004). In addition to the transmission of tacit knowledge, there are also other concepts explaining biotechnology's strong tendency towards spatial clustering including the creation of specialised labour markets (Waters and Lawton-Smith 2008) and business environments (Coenen et al. 2003).

The effect of strong science-industry links and the need for continuous innovation are manifested by the dominance of spin-offs, particularly strong in "red" or medical biotechnology (Landry, Amara and Rherrad 2006). This sub-sector has developed a distinctive model of commercialisation, which is based on DBFs (usually academic spin-offs) refining innovations produced by academic research centres, and selling the products of their R&D to pharmaceutical corporations (Mroczkowski, Khilji and Bernstein 2006, Waters and Lawton-Smith 2008). The spin-offs' main feature is the tendency to establish their business in the vicinity of their parent institution (Abramovsky 2011, Ter Wal and Boschma 2007, Lawton-Smith 2004), which is driven by factors both "hard" – agglomeration economies (Decarolis and Deeds 1999) – and "soft" – participation of entrepreneurs in local social networks related to their work at the parent institution (Sorenson 2003).

Due to high start-up and operating costs (salaries, laboratory equipment) and long R&D pipelines, with the prospect of profitability only after 5 to 7 years of operation (Lawton-Smith 2004), biotechnology firms require external financing. The high risk of R&D failure, which is an integral part of the sector's functioning, limits potential sources of financing to venture capital funds and so-called business angels. The financial factor is also distributed locally due to local market conditions, as venture capital funds usually invest in their close geographic proximity (Lawton-Smith 2004).

Local factors shaping the growth of biotechnology clusters include entrepreneurship culture and the innovative milieu. In the case of industries based on spin-offs, local entrepreneurial culture is crucial, often shaping the attitudes of entrepreneurs and encouraging them (or not) to run their own business (Ter Wal and Boschma 2007). The innovative milieu grows out of a dense network of formal and informal local contacts and knowledge networks. Flows in these networks enable collective learning processes and knowledge spillovers (Lawton-Smith 2004). The intensity of these processes, in turn, affects the thickness of the local institutional environment, which facilitates the establishment of further contacts.

Innovation support policy, shaped by local and regional authorities, has a strong impact on the development of the industry. Support for technology parks, technology transfer units, and collaboration in the science-government-industry triangle, resulting from the concept of the triple helix (Lawton-Smith and Bagchi-Sen 2010), allow the sector to plug into local innovation systems. In many cases, support for the sector, due to erroneous assumptions of creating a new branch of the local economy without taking social and economic conditions into account, turned out to be completely ineffective (Rosiello and Orsenigo 2008). Therefore, the generation of effective strategies and support mechanisms should always be directed toward supporting future solutions with the potential for market success, while taking local social and economic conditions into account (Eliasson 2000).

In addition to the earlier described local factors, global conditions also shape the functioning of the biotechnology sector. This is related to the key role of knowledge, its transmission, and absorption in the functioning and development of the sector. While the transfer of tacit knowledge is usually limited geographically to personal contacts, the transmission of codified knowledge over long distances is not a problem in today's world. Hence, the key role of conferences, trade fairs, and exhibitions as sources of knowledge (especially scientific knowledge) for the biotechnology sector. Furthermore, research staff mobility is also a global phenomenon. As researchers are bonded by cognitive proximity, they are capable of easily joining a variety of research teams (Decarolis and Deeds 1999). In addition, biotechnology companies' contracts with pharmaceutical corporations are global due to strong competition in the marketplace and rapid rate of progress. In effect, pharmaceutical corporations choose the most promising innovation providers, regardless of their geographic location (Mroczkowski, Khilji and Bernstein 2006).

Research on biotechnology clusters (Bagchi-Sen and Scully 2004) has shown that knowledge flows and networks of local and global factors are not competitive but complementary. The complementarity of these types of contacts is explained by the concept of local buzz and the global pipeline (Bathelt, Malmberg and Maskell 2004, Coenen, Moodysson and Asheim 2004). Local buzz includes local, often informal contacts between actors (rumours, news, information). The global pipeline is a set of more formal contacts with the outside world – outside the given geographic cluster. The complementarity of the two types of contacts is derived from the different types of knowledge flowing through them. The development of local contacts causes an inevitable decline in global competitiveness, as most innovation is generated outside, and has to be imported. The neglect of local contacts in favour of global contacts leads to the loss of opportunities for collaboration (such as formation of consortia, joint projects) with local partners (Coenen, Moodysson and Asheim 2004).

The structure and strength of various local and global factors affecting the functioning of companies also shapes their business models. Clarysse, Degroof and Heirman (2003) identify three basic types of companies in the high-tech sector: (1) technological SMEs, (2) prospectors, (3) venture-capital backed companies (Table 11.1).

The technological SME, otherwise referred to as a "lifestyle business" (Raagmaa and Tamm 2004), is the most common form of academic spin-off, managed by researchers and treated as an additional source of income outside of their scientific employment. Most SMEs exist in the form of a sole proprietorship or as a micro-enterprise, and its owners do not abandon their jobs at a parent institution. An important feature of this type of business is its focus on service activity, concentration on the local market, and a lack of plans for expansion and the acquisition of new markets. Often the very fact of generating any profit is satisfactory for the firm's owners (Raagmaa and Tamm 2004).

Table 11.1 Different types of high-tech firms

	Technological SME	Prospectors	Venture capital-backed
Transparency of product/ market	No clear product/ market identified	Still need substantial development of product, market positioning	Clear product/market concept
Capitalisation (EUR)	< 60,000	> 300,000	> 1,000,000
Founding Team	Technical. No hiring of external management	Essentially technical. Some with management experience	Core founding team with technical expertise and external managers
Business Model	Consulting and service-oriented	Product-oriented, product orientation and consulting	Product-oriented
Growth Orientation	Low growth target. Focus on maintaining control over the firm	Growth orientation of the founders, but long period of gestation needed	Exponential growth target built into the business model.
Target Market	Local	Internationally oriented at an early stage	"Born global"

Source: Clarysse, Degroof and Heirman (2003).

The "prospector" type firm is also a spin-off, and the most important difference compared with the technological SME is that the founders are focused on the development of the company and have a defined product that the company can develop and manufacture. Firms are focused on international expansion since their inception, but require a long gestation period to refine their products.

The third type of company is venture capital-backed. In this case, the involvement of external capital and significant capitalisation also allow for a global expansion of the company. Like prospectors, companies backed by venture capital are dominated by manufacturing companies. An important feature in this case is the hiring of external managers, which often leads to conflict with the founders (Mroczkowski and Elms 2009).

Polish Biotechnology Companies – Background

Research for this section was based on an analysis of a database of biotechnology companies in Poland including information on their location, date of establishment, specialisation, and owners' background. The database was supplemented by a series of interviews with entrepreneurs representing Kraków's biotechnology industry cluster. Reports commissioned by the Polish government, local authorities, and industry interest groups as well as cluster initiatives were treated as an additional source of information.

In order to clearly define the biotechnology sector, an OECD classification was used (van Beuzekom and Arundel 2009), and the scope of the research study was limited to dedicated biotechnology firms specialising in R&D. This was due to the need to omit some companies only partially engaged in biotechnology, and specialising in other types of business activity including the production of chemicals and general laboratory equipment.

Post-communist countries in Central and Eastern Europe began to experience the development of a viable biotechnology sector almost 20 years after their Western European counterparts. During the early political and economic transformation period initiated in 1989, the first private biotechnology companies were usually established as spin-offs. However, the institutional framework for the development of Poland's biotechnology industry is still in its infancy. With regard to the eight elements of institutional structure of biotechnology distinguished by Bartholomew (1997), the degree of the commercial orientation of research institutions and the complete absence of a flourishing venture capital market are the weakest points in Poland. The strongest element in this case is Poland's national tradition of scientific education. Poland's changing legal framework, limited role of government in supporting the sector, and expensive patent process cause the performance of Poland's biotechnology sector to lag behind its Western European counterparts.

In early 2012 there were 50 dedicated biotechnology companies operating in Poland, of which 36 were engaged in R&D activities. The sector was geographically concentrated in the largest urban areas including Warsaw, Gdańsk, Poznań, Kraków, Łódź, and Wrocław (Figure 11.1). In addition to "core" DBFs, a number of firms more loosely associated with the sector existed in each of the above cities (Dawidko and Micek 2015), and usually engaged in service and consulting activities.

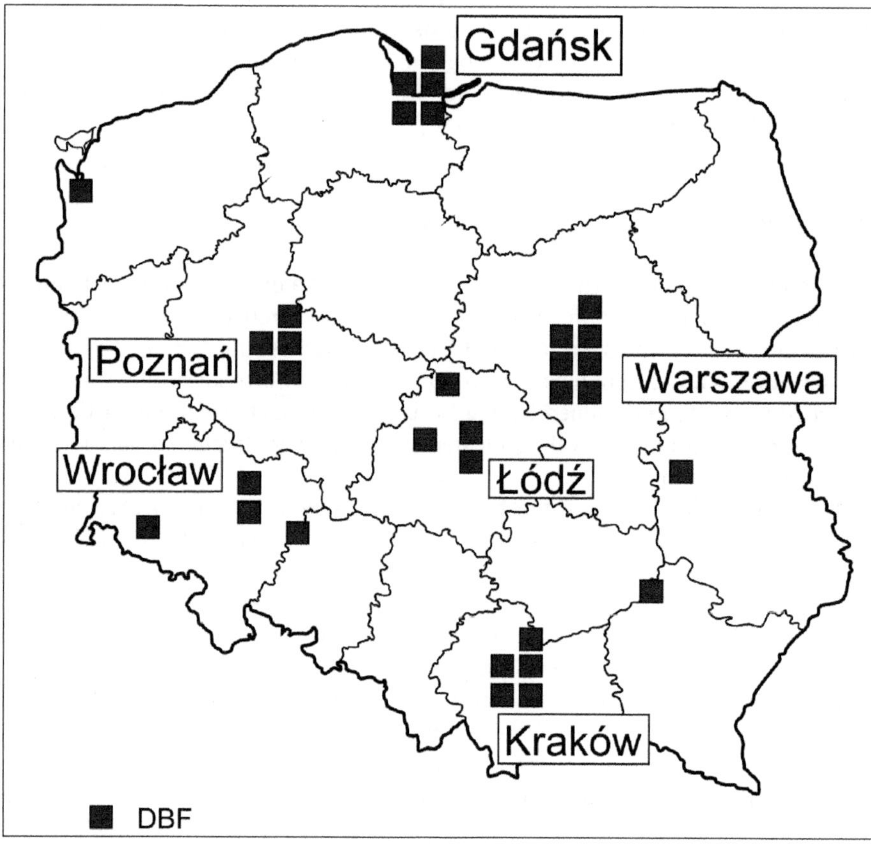

Figure 11.1 Location of R&D dedicated biotechnology firms in Poland
Source: author's own work. As of end of 2011.

The majority of the surveyed companies were small and micro enterprises; only two employed more than 100 people. Total employment in the sector in Poland is estimated at about 3,000 people, of whom 306 are research staff (*Raport* 2007). "Red" is the dominant specialisation in the Polish biotechnology industry. This is related to legal restrictions on the development of agricultural biotechnology including a strict ban on GMO crops and little interest in industrial biotechnology.

Furthermore, the presence of Poland's pharmaceutical industry in the largest cities and well-developed life sciences departments at leading Polish universities constitute the basis of the predominance of medical biotechnology. However, due to financial constraints, the main point of interest of Polish biotechnology companies is the manufacturing of generics, and not innovative pharmaceuticals. Poland's biotechnology sector is also characterised by minimal influx of foreign capital – none of Poland's DBFs are owned by foreign entities or have been established as a result of FDI. Foreign shareholders hold minority shares in some biotechnology companies, but do not have control over any of the surveyed firms.

A total of 22 of the 36 surveyed companies were established as academic spin-offs. Combined with the low influx of foreign capital, this denotes the sector's local origin and confirms previous research conducted in other countries (for example Lawton-Smith and Ho 2006, Lawton-Smith and Bagchi-Sen 2010). The infancy of Poland's venture capital market, constituting the primary source of external financing for the biotechnology sector in developed countries, virtually prevents access to large amounts of start-up capital required for the development of innovative biotechnology companies. Poor access to capital is cited by researchers as the main barrier to the development of biotech spin-offs in Poland (*Raport* 2007). Currently in Poland there exist two venture capital funds dedicated to companies representing the life science industry, which support 11 companies including 2 DBFs. The total lack of activity of foreign capital in this area confirms earlier studies by Lawton-Smith (2004), describing the local preferences of venture capital funds.

In addition, poor connections between the scientific sector and Poland's business community, and the lack of a tradition and culture of entrepreneurship among academic staff, which is a common feature of post-communist countries (Mroczkowski and Elms 2009, Raagmaa and Tamm, Varblane et al. 2007), limit opportunities for academic spin-offs. According to research conducted on behalf of Jagiellonian University (Raport 2007), only 15 per cent of scientists in the life sciences sector employed by leading Polish universities have ever attempted to commercialise the results of their research in any way – not only by setting up their own business, but also through the sale of licenses. Finally, the percentage of researchers running a spin-off firm in Poland is less than one per cent.

Local knowledge flows and spillovers are also key factors driving the development of the biotechnology sector. A characteristic feature of the surveyed companies is minimal commitment to collaboration with other companies. This may be caused by the small number of companies in the sector and their variety – every firm has its own "niche", devoid of competition on the one hand and devoid of partners on the other hand. The functioning of cluster initiatives and business associations varies substantially in terms of overall effectiveness, depending on the location. Business cluster initiatives currently exist in Kraków, Warsaw, and Gdańsk, and attempt to integrate the local business community, academia, and government institutions.

A combination of the above-mentioned factors – academic background, lack of access to venture capital, and an underdeveloped culture of entrepreneurship – creates a situation in which the vast majority of companies function as technological SMEs.

When providing entrepreneurs with access to capital and business knowledge, one may speak of a transformation of companies to prospector or VC-backed companies, depending on the source of capital. Examples of such companies include Bioton – Poland's largest biotechnology company, and Selvita – Poland's first biotechnology company, both of which conduct R&D on innovative pharmaceuticals.

Bioton and Selvita were established as university spin-offs. One reason for their outstanding growth was the commitment of external expertise and capital. Bioton was taken over at an early stage by Ryszard Krauze – one of Poland's business tycoons who invested in the development of the company and its international expansion. One of the founders of Selvita was Pawel Przewiezlikowski, an experienced IT manager, who had acted as a member of the board of Polish companies such as Comarch and Interia.

The business models of the two companies differ significantly. Bioton has become a key manufacturer of recombinant insulin in Europe, controlling nearly a quarter of its market in Poland. However, due to the strength of large pharmaceutical corporations, it was not able to achieve a strong position in the global market despite having made large investments in India and China. Ongoing restructuring is aimed at restricting foreign investment in the unprofitable company.

Selvita's strategy is based on two pillars – its own R&D and business services (research outsourcing). With good income from service activities, the company is able to fund its own research projects focusing on innovative pharmaceuticals. In 2010 Selvita, as the first Polish biotechnology company, signed an agreement with a major pharmaceutical corporation, Orion Pharma, concerning innovative Alzheimer's drug development. Under the agreement, Selvita is responsible for the progress of R&D until the investigated drug is ready for preclinical trials. After this stage, the responsibility for clinical trials, manufacturing, and sale of the drug is transferred to Orion Pharma. Under the agreement, the value of which is estimated at 300 million zł (approximately 100 million USD), Selvita receives tranches of cash after reaching designated milestones in the development of the target drug.

The above suggests that the functioning of a biotechnology business depends largely on local factors – as discussed earlier – including a tradition of entrepreneurship, access to capital, business knowledge, and finally – the quality of scientific research. At the same time, local foundations can be used to build international success, as is the case with the Selvita – Orion business agreement.

An important global factor affecting the functioning of the sector is the international mobility of researchers. According to Dawidko and Micek (2012), Poland is not a country that is attractive to "migrant" scientists. However, Polish scientists often do return to Poland after periods of work in foreign research centres, often representing the world's highest level in their field. Tacit knowledge, experience, and formal and informal contacts established at the time of employment abroad are often used in the conduct of research activity in Poland.

At the same time, global competition is one of the factors limiting the possibility of international expansion of Polish biotechnology companies. Taking into account the small amount of foreign capital flowing into Poland's biotechnology sector, local entrepreneurs can use their knowledge of local market conditions to gain market share in Poland; however, their financial resources are insufficient for international expansion.

Summary and Conclusions

With 50 DBFs employing approximately 3,000 workers, Poland's biotechnology sector is in its early stage of development. As is the case with Western countries, the biotechnology sector in Poland is mainly created by academic spin-offs. However, there exist local factors such as poor access to venture capital and business expertise as well as poor science-industry linkages that constitute a serious obstacle to the development of the sector (Table 11.2). These forces most existing biotechnology businesses to function in the form of technological SMEs or "lifestyle businesses", only as an additional occupation for their owners, often combined with employment at a research or academic institution.

Table 11.2 Global and local factors shaping the Polish biotechnology industry

Global	Local
Specialist knowledge flows	Tacit knowledge flows and spillovers
Partnerships	Institutions
Labour mobility	Entrepreneurial culture
Competition	Financing

Source: author's own work.

Global factors such as the flow of knowledge and personnel may allow for the sharing of expertise, but do not solve any local problems. Most of the surveyed companies in Poland behave in line with their classification (Technological SME) – doing business in local markets, while international expansion is not within the scope of their capabilities. Hence, the concept of local buzz and global pipeline, in the case of Polish biotechnology enterprises, applies only to a select few companies classified as prospector-type (Bioton and Selvita). Other technological SMEs operating in Poland usually do not have any foreign relationships other than scientific contacts (conferences, internships) pursued by their owners who usually work at state universities and other research institutions.

The interaction between global and local factors may have a decisive influence on the development of Poland's biotechnology sector, which is currently in its infancy. The case of Poland highlights the advantages of local factors related to existing political, cultural, and social determinants, as well as the size and model of the given company. The impact of global factors on the technological SME is small, with the exception of specialised knowledge flows. Only the prospect of a company's higher development and transition to a "higher grade" of business activity implies a greater impact of global factors on the functioning of a biotechnology company.

References

Abramovsky, L. and Simpson, H. 2011. Geographic proximity and firm-university innovation linkages: evidence from Great Britain. *Journal of Economic Geography*, 11, 949–77.

Bagchi-Sen, S. and Scully, J.L. 2004. The Canadian environment for innovation and business development in the biotechnology industry: a firm-level analysis. *European Planning Studies* 12(7), 961–82.

Bartholomew, S. 1997. National systems of biotechnology innovation: complex interdependence in the global system. *Journal of International Business Studies* 28(2), 241–67.

Bathelt, H., Malmberg, A. and Maskell, P. 2004. Clusters and knowledge: local buzz, global pipeline and the process of knowledge creation. *Progress in Human Geography*, 28(1), 31–56

Clarysse, B., Degroof, J.J. and Heirman, A. 2003. *Growth Paths of Technology-based Companies in Life Sciences and Information Technology*. Luxembourg: Office for Official Publications of European Communities.

Coenen, L., Moodysson, J., Asheim, B.T. and Jonsson, O. 2003. *The Role of Proximities in Knowledge Dynamics in a Cross-Border Region: Biotechnology in Oresund*. Copenhagen: DRUID Summer Conference.

Coenen, L., Moodysson, J. and Asheim, B.T. 2004. Nodes, networks and proximities: on the knowledge dynamics of the Medicon Valley biotech cluster. *European Planning Studies*, 12(7), 1003–18.

Cooke, P. 2004. The accelerating evolution of biotech clusters. *European Planning Studies*, 12(7), 915–20.

Dawidko, P. and Micek, G. 2012. Labour flows in the biotech sector in Poland. *Prace Geograficzne Jagiellonian University*, 130, 73–89.

Dawidko, P. and Micek, G. 2015, The spatial evolution of the Polish biotech industry: a path-dependent process? *European Planning Studies*, 23(5), 944–962.

Decarolis, D.M. and Deeds, D.L. 1999. The impact of stocks and flows of organizational knowledge on firm performance: an empirical investigation of the biotechnology industry. *Strategic Management Journal*, 20(10), 953–68.

Eliasson, G. 2000, Industrial policy, competence blocs and the role of science in economic development. *Journal of Evolutionary Economics*, 10, 217–41.

Geenhuizen, M. and Soetanto, D. 2008. Science parks: what they are and how they need to be evaluated. *International Journal of Foresight and Innovation Policy*, 4(1/2), 90–111.

Gertler, M.S. 2003. Tacit knowledge and the economic geography of context, or the undefinable tacitness of being (there). *Journal of Economic Geography*, 3, 75–99.

Gertler, M.S. and Vinodrai, T. 2009. Life sciences and regional innovation: one path or many? *European Planning Studies* 17(2), 235 61.

Landry, R., Amara, N. and Rherrad, I. 2006. Why are some university researchers more likely to create spin-offs than others? Evidence from Canadian universities. *Research Policy*, 35(10), 1599–1615.

Lawton-Smith, H. 2004. The biotechnology industry in Oxfordshire: enterprise and innovation. *European Planning Studies*, 12(7), 985–1001.

Lawton-Smith, H. and Ho, K. 2006. Measuring the performance of Oxford University, Oxford Brookes University and the government laboratories' spin-off companies. *Research Policy*, 35(10), 1554–68.

Lawton-Smith, H. and Bagchi-Sen, S. 2010. Triple helix and regional development: a perspective from Oxfordshire in the UK. *Technology Analysis & Strategic Management,* 22(7), 805–18.

Malo, S. and Norus, J. 2009. Growth Dynamics of dedicated biotechnology firms in transition economies. Evidence from the Baltic countries and Poland. *Entrepreneurship & Regional Development*, 21(5–6), 481–502.

Mroczkowski, T., Khilji, S.E. and Bernstein, B. 2006. From invention to innovation: toward developing an integrated innovation model for biotech firms. *Journal of Product Innovation Management*, 23, 528–40.

Mroczkowski, T. and Elms, H. 2009. Tracking progress: two approaches to biotechnology development – cases from Central Europe. *Journal of Commercial Biotechnology*, 15, 227–35.

Raagmaa, G. and Tamm, P. 2004. An emerging biomedical business in a low capitalised country. *European Planning Studies*, 12(7), 943–60.

Raport nt. barier komercjalizacji wyników badań naukowych w dziedzinie life science w Małopolsce, 2007, Kraków: Jagiellońskie Centrum Innowacji.

Rosiello, A. and Orsenigo, L. 2008. A critical assessment of regional innovation policy in pharmaceutical biotechnology. *European Planning Studies*, 16(3), 338–57.

Sorenson, O. 2004. Social networks and industrial geography. *Journal of Evolutionary Economics*, 13, 513–27.

Ter Wal, A.L.J. and Boschma, R.A. 2007. Co-evolution of firms, industries and networks in space. *Papers in Evolutionary Economic Geography*, 07/07.

Van Beuzekom, B. and Arundel, A. 2009. *OECD Biotechnology Statistics*. Paris: OECD.

Varblane, U., Dyker, D., Tamm, D. and Tunzelmann, N. 2007. Can the national innovation systems of the new EU Member States be improved? *Post-Communist Economies*, 19(4), 399–416.

Waters, R. and Lawton-Smith, H. 2008. Social networks in high technology local economies: the case of Oxfordshire and Cambridgeshire. *European Urban and Regional Studies*, 15(1), 21–37.

Chapter 12

Agricultural Tourism Farms in Poland: How the Farmers Improve their Businesses—A Case Study

Sławomir Dorocki, Anna Irena Szymańska,
Małgorzata Zdon-Korzeniowska

Introduction

This chapter discusses the actions taken by Polish farmers in the realm of agricultural tourism, especially the commercial aspects of this form of tourism. The role of innovation is analyzed, as are various aspects of family business operations and local business operations. The chapter is largely based on survey research conducted in early 2012, with the participants being Polish agricultural tourism enterprises. The questionnaires were sent to 1,326 agricultural businesses all over Poland. Finally 46 agriculture enterprises took part in the questionnaire survey. Another key part of the chapter consists of a case study of Na Zagrodzie Tourist Farm located near the village of Lubień in Małopolska Province (Polish: Voivodeship). Other tourist farms are also discussed in the chapter.

The analysis of Polish agricultural tourism farms is rooted in the research literature. The concepts of agricultural tourism and agricultural tourism farms have been defined by Drzewiecki (2001, 2009) and Sznajder and Przezbórska (2006). In order to enhance the definitions mentioned above, legal documents (Ustawa 2004) were analyzed that regulate business activity in Poland. Tax laws (Ustawa 1991) were also analyzed. The research literature was used to identify key drivers of business growth for agricultural tourism farms in Poland. This included important internal drivers (farm-dependent) and external drivers (farm-independent) (Van Der Ploeg et al. 2000, Wiatrak 2003, Rachwał 2005a, 2005b, Zioło 2007, Kurtyka 2008, Knecht 2009, Palicha 2009, Wojciechowska 2009, Zdon-Korzeniowska 2009, Phillip, Hunter and Blackstock 2010, Borowiec and Rachwał 2011, Dorocki, Szymańska and Zdon-Korzeniowska, 2012a, 2012b, Sikora 2012, Dorocki et al. 2012, Dorocki and Zdon-Korzeniowska 2013).

Concept of Agricultural Tourism

The concept of agricultural tourism is understood in a number of ways in the research literature and in agricultural practice itself. Often agricultural tourism is a synonym of rural tourism or tourism in rural areas. The research literature defines agricultural tourism as a form of rural tourism, which is fundamentally based on tourism in rural areas (Marcotte, Bourdeau and Doyon 2006, Mazurski 2010). While a deeper analysis of the definition of agricultural tourism is possible, the basic idea behind agricultural tourism is based on the notion that the provider of tourist services is a farm or rural areas in general (Sznajder and Przezbórska 2006). A universal definition of present-day agricultural tourism is provided by Drzewiecki (2009: 13) who formulates it in the following manner: "Agricultural tourism is a form of recreation taking place in rural areas characterized by agriculture and it is based on accommodations and recreational activity at a farm or equivalent and its natural, production and service environment."

However, this definition and its practical implications do not encompass the entire spectrum of variety and changes in the development of tourist services in rural areas. The same is true of legal aspects, especially those associated with laws on the freedom to pursue business activity and income tax laws, all of which often serve as sources of abuse in the business of agricultural tourism.

While it may be difficult to produce an unequivocal definition of agricultural tourism and identify which entities are involved in the business of agricultural tourism, it is clear that the offering of agricultural tourism services is a leading form of non-agricultural business activity in rural areas in Poland. Agricultural tourism is also a key pillar of the strategy of multifunctional rural development (Pimbert et al. 2001). According to Sikora (2012: 46), "It is commonly accepted that multifunctional rural development consists of the diversification of economic activity in rural areas, which includes the initiation of non-agricultural forms of economic activity." The European Union appears to recognize the role and significance of agricultural tourism in the life of rural areas. Its stance on tourism in rural areas and agricultural tourism is quite clear in that "rural tourism constitutes a fundamental element of rural development policy and is a very important and complementary part of rural development strategy, and furthermore, plays a special role in spatial planning" (Sznajder and Przezbórska 2006: 53).

A review of terminology with respect to the legal aspects of agricultural tourism also appears to be important. Agricultural tourism—as a business—is governed by Poland's *Act on the freedom to pursue business activity* (Ustawa 2004) as well as the *Act on income tax* (Ustawa 1991). Both acts of law define what it means to be an agricultural tourism business. Poland's tax law defines agricultural tourism as an activity that supplements income from agriculture. Poland's personal income tax law (Article 21, Item 43) states that personal income may not be subject to taxation if it is generated by the renting of rooms to visitors who meet the following requirements:

- rooms are rented to individuals visiting for recreational purposes
- rented rooms are located in residential buildings
- landlord operates a farm in a rural area
- rented rooms are located in buildings found on landlord's farm, and
- number of rented bedrooms does not exceed five.

In addition, income generated by the food services provided by the landlord to the visitors is also not taxable (Kapała 2010).

Furthermore, the *Act on the freedom to pursue business activity* does not classify the following as non-agricultural activity:

- agricultural production
- renting of rooms and places to pitch a tent to visitors
- sale of homemade foods, and
- rendering of other tourist-related services by farm operators.

In effect, agricultural tourism farms need not be registered as business entities (Ustawa 1991). In summary, the laws just described make it legal to not register an agricultural tourism farm as a business if the number of rooms rented to tourists does not exceed five. This also means that this type of farm does not need to pay income tax on its tourist activity. However, in reality, some regular farms employ deceptive tactics to avoid paying income taxes by relying on the above tax exemption. In other cases, farmers are only farmers by definition thanks to a variety of deceptive ownership practices (Hugues 2004, Drzewiecki 2009). The lack of clarity in terms of definitions and tax laws make it possible for a variety of tourist-type businesses to exist in rural areas in Poland.

In general it may be assumed that entities offering agricultural tourism services in Poland function within the framework of regular farms or in the form of self-employment based on the freedom to pursue business activity, various income tax laws, and various legal forms of employment. The latter is essentially a form of business. The former—agricultural tourism on a working farm—may also be considered a business, as farmers need to look at their daily work from a market perspective. This includes a consideration of customer needs, identification and tracking of competitors, preparation of a competitive offering, marketing and pricing. In summary, agricultural tourism services are a form of business activity, regardless of whether this business is registered as a business or exists within the framework of a working farm. Ultimately, the basic purpose of this type of activity is profit (Mamdy, Guillot and Disez 2007, Sznajder, Przezbórska 2006).

Growth Factors in the Agricultural Tourism Industry

In an attempt to take a broader view of agricultural tourism and the determinants of its growth, it is necessary to identify two basic areas of analysis. The first is external factors that cannot be controlled by farm operators to any great extent. The second is internal factors that can be controlled by farm operators. External factors include the following (Rachwał and Zdon-Korzeniowska 2011, Dorocki et al. 2012):

- Domestic factors—changes in the structure and size of the domestic farm sector, farm profitability, ability to offer additional goods and services (for example homemade wine), cultural and environmental attractiveness of given farm location, domestic social and economic situation, level of tourist infrastructure development, quality of tourist infrastructure, support from organizations promoting the development of agricultural tourism, tax laws, organizational and legal solutions associated with business development and agricultural production, state of the natural environment.
- Changes in life needs, preferences, and lifestyles of contemporary "information society"—ways of spending free time, ways of planning recreational activities, new trends in travel destinations, choice of tourist goods and services.
- Global economic crisis resulting in lower potential customer incomes, leading to lower expenditures on tourist attractions over a relatively short period of time.
- Domestic and foreign competition from other forms of tourism.

Geographic location is a key external determinant of success in the agricultural tourism sector. Its location relative to major cities, transportation networks, as well as other tourist attractions (for example tourist hotspots, winter resorts, historic towns, nature reserves) is also very important. Another key factor is landscape—mountain areas, coastal areas, lake districts, forests. The quality of the natural environment is also relevant (Pałka 2008). The immediate vicinity of an agricultural tourism farm is likewise important, as it directly affects the quality of the tourist experience. The best locations are isolated locations found away from urban and industrial areas. Location quality can be estimated in terms of low population density, small percentage of non-agricultural jobs, large number of traditional farms using open structures, diversified agriculture (crops and animals), large number of medium size farms, generally non-intensive agriculture and forest management. In addition, agricultural tourism regions are characterized by a homogeneous cropland landscape and low income, which may motivate some residents to pursue non-agricultural sources of income (Drzewiecki 2001, Kubal and Mika 2012).

The owners of agricultural tourism farms have little capability to influence external factors; however, research has shown that many owners make a joint effort to do just that in the form of farm owners' associations and collaboration with public organizations in order to improve their chances of success (Wiatrak 2003).

At the unit level, the most important factors driving agricultural tourism growth and operational stability are human resources, specialization of the given farm, and the quantity of fixed assets and their quality.

Research has shown that the key to success in agricultural tourism appears to be human resources. The human capital derived from entire families of farmers determines how well their business enterprise functions and creates a specific ambience that affects the tourist experience. However, the owners and employees of an agricultural tourism farm need to have certain entrepreneurial, psychological, intellectual and personal predispositions in order to be successful (Wojciechowska 2010). Good physical condition is also an important precondition. The above characteristics are needed of all members of the family in contact with tourists. Owners and other farm workers need to be able to express themselves—often in foreign languages. Their ability to meet a full spectrum of tourist needs is also of the utmost importance. Some knowledge of basic psychology is also helpful—as some guests may have a very different personality or expectations.

The following personal characteristics are found in successful entrepreneurs: 1) being active, 2) determination, 3) strong work ethic, 4) ability to take the initiative, 5) creativity, 6) assertiveness, 7) confidence, 8) belief in one's ability to succeed, 9) self-discipline, 10) integrity, 11) willingness to take a calculated risk, and 12) being responsible for oneself and others (Rachwał 2012). Additional characteristics, which tend to be helpful, include the ability to identify and take advantage of opportunities in the market, ability to map and pursue a self-established career path, ability to work in a team, ability to plan and execute a plan, ability to communicate with others, and the ability to manage stress. Finally, an effective entrepreneur is not afraid of difficulties and new challenges. He or she remains open to the world, acquires new skills and knowledge, and follows a code of ethics (Rachwał 2005a, 2005b, Borowiec and Rachwał 2011).

Other important internal factors include the quality of fixed assets such as buildings, open structures, and machines. The aesthetic value of an agricultural tourism farm is also becoming increasingly important. A farmer who decides to pursue agricultural tourism needs to manage a large number of variables, which include creative and innovative marketing, farm aesthetics, event management, time management, team management, farm work management, planning work, market analysis and customer service (Gale 1997). In summary, an entrepreneur in the modern agricultural tourism sector needs to be competent in both the economic realm and the psychological realm.

Agricultural tourism is the most popular type of non-agricultural business activity in rural areas. It also appears to be a very desirable form of rural business activity. Agricultural tourism is based on existing land and human resources including family members and their time, farm houses, farm buildings, farm machines, animals, local natural environment, local knowledge, skills, and work effort. All of the above yield less investment risk. Business activity is essentially an expansion of existing agricultural activity. It is not a start-up that requires significant capital and work input. The provision of agricultural tourist services takes place on a "cease and relate" basis. Farm owners and employees render services related to agricultural work and household management—things they already know how to do.

Farm owners and employees are more confident in their new business venture, as they already know how to run a farm and household. Agricultural tourism usually does not require significant capital. Some rooms may need to be adapted or modernized for tourist use. Farm owners and employees as well as tourists benefit from a mutual exchange of ideas. The arrival of new tourists every year brings the world closer to rural areas. Tourists learn about rural life and rural residents learn about non-rural life. The continuation of actual agricultural work may also be of some benefit, as many farmers believe that agriculture has special value. In this case, the relative unprofitability of agriculture can be offset by the profitable nature of agricultural tourism. Furthermore, tourists may perceive agriculture as a tourist attraction.

Agricultural tourism also creates an opportunity to sell farm products. The farm owner and other local farmers may entice tourists to buy their products. Agricultural tourism also creates an opportunity for rural residents to become active in business and may enhance rural pride. This type of economic activity helps maintain material and cultural heritage in rural areas. Multiplier effects can also benefit the area as a whole, assuming that the negative effects of mass tourism do not offset the positive effects. Large infrastructural investments can sometimes alter the rural character of rural areas (Kurtyka 2008). Tourist expenditures remain in the local economy. Agricultural tourism farms and other entities in the area as well as local governments all benefit financially.

A number of tourists have become interested in so-called alternative tourism in recent years. Some tourists wish to escape the noise of the city and return to their roots. This appears to be a byproduct of urbanization. All of the above factors help in the development of rural-based businesses. This is important, as rural areas are most affected by global changes and changes within the national economy. Rural areas tend to experience more unemployment, aging of society, depopulation, social apathy and economic apathy.

Drzewiecki underscores the important role of agricultural tourism in the modernization of rural areas, noting that agricultural tourism fits the ecological model of sustainable human and agricultural development (Drzewiecki 2001). This conclusion suggests that agricultural tourism is desirable and is bound to grow in Poland and in the world.

Growth of Agricultural Tourism in Poland

The first step towards tourism in Polish rural areas can be traced back to Tytus Chałubiński who promoted tourist stays at the homes of Podhale Highlanders. Tourist stays in rural areas adopted an organized format in the 1930s (Drzewiecki 2001). The organization of tourist trips and the adaptation of rural environments to tourist needs were first initiated by local governments, community organizations and economic associations. This included the Gromada Tourist Cooperative established in 1937. In 1957 Poland's government made it legal to create "vacations under a pear tree" on farms (Drzewiecki 2001). Demand for tourist and recreational services sparked an increase in the number of tour operators in Poland, including the Workers' Vacation Fund. The number of towns offering tourist attractions increased as well. Some towns offered an attractive natural environment. Some offered close geographic proximity. Others offered authentic agricultural traditions (Drzewiecki 2001).

In the mid-1970s, Poland's government began to require Provincial Tourist Boards to help tourists find accommodations. This led to a decrease in tourist traffic in rural areas. The following years were not good for rural tourism due to political and economic factors present in Poland at the time (Drzewiecki 2001). This changed in the 1990s when Polish farmers acquired the right to rent rooms to tourists the same way a hotel would (Sikora 2012).

The growth of the agricultural tourism sector in Poland was aided by the establishment of various local, regional, and national organizations and associations focused on agricultural tourism. These organizations promoted rural tourism, and at the same time, prompted rural businesses to increase the quality of their facilities and services. The largest and most important organization working to increase the popularity of agricultural tourism in Poland is the Polish Federation of Rural Tourism, also known as "Hospitable Farm" (established in 1996). It is a national non-profit that consists of 45 local and regional associations (Agroturystyka.pl 2012). It is an association of associations that ranks rural lodging facilities using its WBN database, which was created and implemented as part of the TOURIN II program known as "Rural tourism in rural areas and forests." The program's criteria mirror similar ranking programs in other countries in the European Union. The WBN system ranks participating facilities based on overall quality and services offered.

Poland's economy was affected by an array of restructuring efforts in the 1990s, which forced many social, economic, and spatial structures to adapt to the new rules of the game (Zioło 2007). This adaptation process proceeded unevenly across Poland and resulted in large differences in economic development across the countryside (Chrapek 2010).

According to Poland's Ministry of Agriculture and Rural Development, there are about 12,000 agricultural tourism farms in Poland featuring over 180,000 beds. The number of tourists visiting rural areas in Poland ranges from 1.0 to 1.5 mln per year. However, the Polish Federation of Rural Tourism notes that accurate data on the number of tourist farms does not exist. It is assumed that about 10 percent of tourist farms are registered, while 90 percent operate illegally. Hence, unregistered farms are strong competitors and may lower prices and increase profit at the same time.

The type of agricultural tourism farms varies across Poland, as does their type of agricultural production (Figure 12.1). Large commercial-type agricultural tourism farms can be found across western and northeastern Poland, while the central and southern part of the country is characterized by smaller family farms. Eastern Poland features extensive farming (Wojciechowska 2009). Farms in coastal Poland and southern Poland are difficult to classify in that many do not really fit the concept of rural tourism but rather simply offer rooms for rent, which would make them urban-type lodging facilities. In general, the areas most eager to embrace agricultural tourism are low-income areas with a pristine natural environment.

The literature and statistical data collected in the research process suggest that about 40 percent of all agricultural tourism farms in Poland are located in three voivodeships: Małopolskie, Podkarpackie, Warmińsko-Mazurskie (Turystyka wiejska 2012).

The business behaviors of agricultural tourism farms were evaluated via direct survey research. The high degree of infrastructural development at agricultural tourism farms and the wide variety of services offered to tourists suggest a very high degree of organizational and financial effort on the part of agricultural tourism farms. The extensive tourist offering at many agricultural tourism farms (over 30 percent) is aimed at everyone interested in this particular type of recreation. Some farms are more specialized. This includes farms that cater to children (19.7 percent), company employees (12.3 percent), foreign tourists (10.7 percent), and senior citizens (9.8 percent). Most farm operators are also able to speak at least one foreign language: English—47.8 percent, German—41.3 percent, Russian—43.5 percent, Italian—4.3 percent, Ukrainian—4.3 percent, Spanish—2.1 percent).

Farm owners participating in the survey noted that innovation plays an important role in agricultural tourism farm management. Almost 63 percent of survey participants noted the introduction of some type of innovation last year. The three key sources of innovation were guest feedback (36.4 percent), strategies used by competitors (24.2 percent), and ideas from family members (24.2 percent).

The surveyed farms tended to use information technology as part of their business operations. This was especially true of the Internet. Almost 93 percent of the farms surveyed had Internet access—83 percent of which was wireless. Online reservations are available at over 63 percent of the farms surveyed. Many of the farms go beyond having an e-mail address and a website, and use the Internet for marketing purposes. This includes having a profile on social media (21.7 percent), web banners and ads on local community websites (15.2 percent), and e-mail advertising (6.5 percent). Each type of marketing strategy above indicates the willingness of farms to pursue innovative avenues of business promotion.

Description of "Na Zagrodzie" Agricultural Tourism Farm

The next part of the chapter consists of a case study of the "Na Zagrodzie" agricultural tourism farm. The purpose of the case study is to evaluate the business behavior of the farm's owners.

The "Na Zagrodzie" agricultural tourism farm is located in the village of Lubień and is operated by Mr. and Mrs. Wojtan. It is a multigenerational farm currently occupied by seven individuals: the owners, three children, owner's father and sister. The owners emphasized that a multigenerational farm lowers operating costs and creates a specific family atmosphere.

The village of Lubień has about 3,000 residents and is located in Małopolskie Voivodeship in southern Poland. The area is mountainous and attractive to tourists. It is part of the Beskid Wyspowy Range in the larger Carpathian mountain chain. A characteristic feature of the area is the presence of lone peaks that create an interesting landscape. The village is located in the Raba Valley and is surrounded by mountain peaks popular with tourists: Szczebel, Luboń Wielki, Zębalowa. Several tourist trails pass through the village. Two of the trails lead to popular tourist sites. However, it was not the natural environment, but the fact that the village is located along a major road in the valley that has made the village a good lodging area. The village of Lubień has served as a lodging area since the early-nineteenth century when travelers going from Kraków to the Podhale region would stop overnight (Figure 12.1). The village is also located close to two regional resort towns—Rabka-Zdrój and Mszana Dolna. Lubień is in itself a popular tourist spot with vacation facilities and a large number of restaurants for a village of its size. Nine registered agricultural tourism farms operate in the Lubień area.

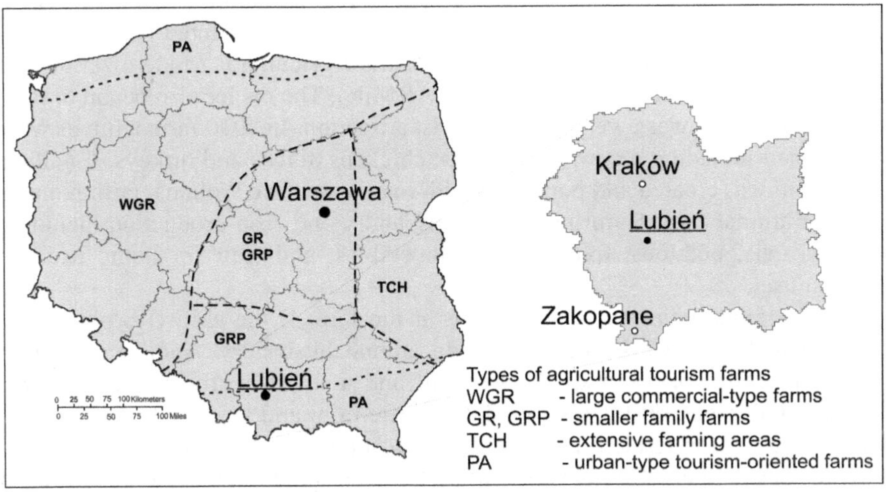

Figure 12.1 Types of agricultural tourism farms in Poland and geographic location of the village of Lubień

Source: authors' own work.

The agricultural tourism farm operated by the Wojtans is a typical small farm of about two hectares. Farmland in the Lubień area has been significantly subdivided over the years. The soil in the area is also not very good. Both factors have prompted local farmers to pursue non-agricultural sources of income. Local farmers supplement what they produce on their farms with other forms of employment. Most farm products in the area are destined for home use. This includes potatoes (major crop) and plants destined for farm animals. Most farms only produce what the farmer's family and their livestock can consume. However, the Wojtans' farm possesses the farm equipment needed to intensify agricultural production if economic conditions change or more land can be leased. Hence, the Wojtans' farm is a typical example of a farm with diversified agricultural production, which is rooted in a long tradition of farming. This affects the nature of its agricultural tourism business and indicates that the farm is shifting towards non-agricultural uses.

The Wojtans' farm was already accommodating tourists in the 1970s. While this was not a formal lodging operation, the visitors were steady customers who came year after year. The services rendered by the Wojtans at the time were limited to lodging and the provision of food. Tourists came for the quiet rural atmosphere and the low price. The farm began to operate as a formal tourist destination in 2003 when Poland's economy began to slow down and the job market became difficult. The owner of the farm lost his job in the non-agricultural sector and needed a new source of income. The farm had to be adapted for use by tourists based on existing laws and regulation in the tourist sector. Rooms in the house and in other buildings needed to be adapted for tourist use. This was accomplished using funds from the Agricultural Modernization and Development Agency (Polish acronym: ARMiR). The owner also began to raise ostriches, which was a very unique tourist attraction. In 2010 farm animals were purchased including a cow, sheep, pigs, chickens, ducks and turkeys. Finally, a fish pond was created and populated with royal carp. The Wojtans' farm remains an operational farm featuring animal husbandry and crop production including barley, oats, potatoes, fodder beets, vegetables, and fruit for home use and tourist use.

One of the main tourist attractions at the farm is the ability to participate in seasonal farm work and year-round animal husbandry. Another attraction is the presence of a working water mill and a small hydroelectric plant. An antique buhrstone mill is also available at the farm and was used until recently to mill grain for fodder. The water mill features a water wheel, as does the hydroelectric plant. The farm is waiting to be certified as an ecological farm. It offers homemade jams and smoked meats. One of its specialties is homemade sauce. The farm offers a large array of meeting opportunities in a natural setting. Visitors can enjoy a grill parlor, gazebo, outside areas with a roof, large yard with playground, sandbox, volleyball court, and other attractions. Five rooms are available to tourists, which can accommodate 16 people at a time. Wireless internet access is also available. Handicapped visitors can also be accommodated. The owners are open to visitor requests such as special events and corporate training sessions.

Na Zagrodzie Agricultural Tourism Farm as an Example of an Innovative Family Business

The research on Na Zagrodzie agricultural tourism farm was performed based on in-depth interviews, which covered both traditional business ideas and the notion of innovation in business. The owners of the farm were interviewed as experts and practitioners in the field. The interview questions were divided into three categories:

1. To what degree is this a family farm?
2. What types of innovative business practices are employed at the farm?
3. What business models are used at the farm?

Agricultural tourism farms are first and foremost family businesses due to the nature of the business as well as the role of cultural heritage. The seasonal nature of farm work and its dependence on weather make it necessary to employ the services of all family members. According to the owner of Na Zagrodzie, all members of his family take part in farm work and in the process of making important strategic decisions. This includes children to the best of their ability. One of the key aspects of the operation of a farm is the division of labor between the owners. The owner's wife makes decisions regarding interior design and runs the kitchen. The owner decides what crops to plant and when to plant them. He also takes care of administrative issues as well as farm machines and the needs of farm animals.

The Na Zagrodzie agricultural tourism farm is continuing a tradition of rendering tourist services, a tradition handed down from one generation to the next. The owner of the farm perceives the family nature of the farm as an asset. He believes that this nature helps the business grow and helps the family grow as well. He lists the following as being important:

- Self-sufficiency linked to the right division of labor among family members.
- The division of labor unites the family. It increases the family's sense of trust, safety and responsibility. This responsibility also includes the home and other farm property.
- Minor conflicts between family members teach the art of compromise and the ability to engage in inter-generational dialogue.

A characteristic trait of family businesses is the planning process involved with the inheritance of the business by the next generation. The inheritance of the business is more sensible when the next generation works along with the current owners. This helps the future owners better understand the unique aspects of the business via participation in key business decisions. It is important for the current owners to educate the future owners and teach them a variety of specific skills. The owners of Na Zagrodzie make an effort to teach their children the art of running an agricultural tourism business, which includes product development and high quality customer service. Other important skills include the ability to listen to customer requests as well as communicate with customers in a variety of ways and languages.

The data collected on the farm suggest that the farm is quite innovative. The offering is rather original with its ostrich feature and it is being enhanced with the purchase of other animals. Other innovative elements include special events and corporate events. The owner believes that innovation is a key part of the farm's growth strategy, which he defines as "the ability to manage the business, adopt new technologies, and create new products and business opportunities in order to make a profit." He perceives himself to be an entrepreneur who is innovative by "creating something new or improving something existing." The owners point to the following to show their ability to innovate:

- Customers—"listening to customers who articulate their expectations"
- Competitors—"studying what other farms are doing via the internet and trade fairs"
- Oneself and family members—"coming up with new self-generated ideas"

The Na Zagrodzie agricultural tourism farm makes an effort to inform and entice potential customers. The following marketing tools are employed by Na Zagrodzie:

- brochures, folders
- website and online ads
- ads posted across the township
- marketing campaigns—for example half-price weekends—Discover Beskid Wyspowy campaign
- participation in promotional events such as the Agricultural Tourism Fair run by the Agricultural University of Kraków, a booth at the County Harvest Festival, various township events, and participation in the Polagra Farm Fair in Poznań in 2005
- Presence across social media—promotional clip in YouTube
- Participation in television programs such as "The Egg" (TV Kraków, TV Polonia), which used images of Na Zagrodzie; live television program run by TV Kraków that discussed Easter traditions and egg painting at Na Zagrodzie in 2009.

The marketing identity of the farm is an important part of the business. Na Zagrodzie has its own logo with an African ostrich and a water wheel in the background.

The owners of Na Zagrodzie plan to create educational opportunities at the farm. Such educational farms already exist in France. Students could earn educational credits at the farm by observing the production of bread and milk products (cheese, butter, cream, yogurt). The farm already has a watermill. Today preschoolers and elementary school students visit the farm to learn about agricultural production and to visit the farm's petting zoo. In addition, the farm offers an enhanced tourist offering (Kmita-Dziasek 2009) by providing onsite food service and homemade food products for later use. The owner's wife also teaches cooking to tourists staying at the farm. This is especially appealing to foreign tourists.

The farm maintains a guestbook, currently with 143 signatures, 17 of which belong to guests from Austria, Italy, France, England, Finland, Scotland, and Australia. The first entry comes from September of 2003. The farm has had 126 visits from domestic tourists, mainly from Kraków and other larger Polish cities such as Gdańsk, Kołobrzeg, Warszawa, Radom, Częstochowa and Wrocław.

Na Zagrodzie Farm won third place in the agricultural tourism category in 2007 as part of the eighth regional tourism campaign called "Discovering Małopolska Voivodeship." The award helped confirm that the farm's owners are on the right track towards meeting customer expectations.

The main barrier to further growth is financial limitations. The economic situation is such that the farm's owner receives KRUS health benefits and leases land. This makes him ineligible for European Union subsidies for further growth. One solution to the growth problem would be more foreign tourists for whom this type of recreation is very affordable.

Conclusions

Research has shown that agricultural tourism farms require certain specific types of entrepreneurial behaviors of farm owners in order for such a business to be successful. This type of business does offer a chance for farm owners who find traditional farming to be less than profitable. This is especially true of small farms. The linkage between agricultural production and tourist services helps enhance farm profitability, promote entrepreneurial behaviors, stir innovation, and enrich the tourist offering at the same time. Agricultural tourism farms also help rural areas maintain traditional professions and cultural heritage.

Hence, agricultural tourism farms not only affect the economics of rural areas, but also the mentality of rural residents. Owners of agricultural tourism farms are a unique brand of entrepreneur who hands down his or her business to future generations. This helps maintain an entrepreneurial spirit in rural areas across generations. Farmers are forced to make key decisions, calculate profitability, and follow general economic trends. This type of thinking has been around for a very long time in areas with mostly private farms. However, areas of Poland with primarily former state-owned agricultural enterprises tend to exhibit less of an entrepreneurial spirit. One example of this is the Warmia region in northern Poland. In this part of Poland, many agricultural tourism farms were started by outsiders and not local farmers. A similar situation exists in highly attractive tourist areas where outside investors compete with local farmers for business.

It appears that a proactive entrepreneurial approach to agricultural tourism is needed for growth to occur in rural areas. Other key ingredients include best practices and models. For example, the owner of Na Zagrodzie noted that his visit to French rural areas played an important role in the growth of his business. The exchange of ideas helped his business grow. While the nature of agricultural production in France significantly differs from that in Poland, the lessons of French agricultural tourism farms can still affect decisions made by similar farms in Poland. Na Zagrodzie is now a role model in itself. Its business model helps shape the entrepreneurial views of other local farmers. One proof of this is the collaboration between local farmers who wish to enhance tourist services in the township. One potential inference in this case may be that the entrepreneurial practices of one farmer may affect those of other local and non-local farmers.

References

Agroturystyka.pl, Available at: www.agroturystyka.pl [accessed: 27 June 2012].

Borowiec, M. and Rachwał, T. 2011. Kształtowanie postaw przedsiębiorczych na lekcjach geografii wyzwaniem edukacyjnym w procesach globalizacji, in *Przedsiębiorczość w warunkach globalizacji*, edited by Z. Zioło and T. Rachwał, series: Przedsiębiorczość-Edukacja, 7, Warszawa—Kraków, 321–32.

Chrapek, G. 2010. *Agroturystyka na Podkarpaciu, Turystyka i rekreacja.* Państwowa Wyższa Szkoła Techniczno-Ekonomiczna w Jarosławiu, 1, 5–20.

Dorocki, S., Rachwał, T., Szymańska A.I. and Zdon-Korzeniowska, M. 2012. Spatial conditions for agritourism development on the example of Poland and France. *Current Issues in Tourism Research*, 2(2), 20–29.

Dorocki, S., Szymańska, A.I. and Zdon-Korzeniowska, M. 2012a. *Polskie gospodarstwa agroturystyczne jako przedsiębiorstwa rodzinne,* in *Firmy rodzinne—współczesne wyzwania przedsiębiorczości rodzinnej. Determinanty rozwoju*, edited by A. Marjański, series: Przedsiębiorczość i Zarządzanie, 13(8), Łódź: Społeczna Akademia Nauk, 45–60.

Dorocki, S., Szymańska, A.I. and Zdon-Korzeniowska M. 2012b. *Agritouristic Family Enterprises in Poland,* paper to the conference 'Entrepreneurship, Family Firms and Economic Development,' Kraków, April 27–28, 2012 (in print).

Dorocki, S. and Zdon-Korzeniowska, M. 2013. Rozwój agroturystyki jako formy dywersyfikacji działalności rolniczej—przykład Francji i Polski, in *Planowanie rozwoju przestrzeni wiejskiej,* edited by K. Kurowska, Urząd Marszałkowski Województwa Warmińsko-Mazurskiego, 44–68.

Drzewiecki, M. 2001. *Podstawy agroturystyki.* Oficyna Wydawnicza Ośrodka Postępu Organizacyjnego, Bydgoszcz.

Drzewiecki, M., 2009. *Agroturystyka współczesna w Polsce.* Gdańsk: Wyższa Szkoła Turystyki i Hotelarstwa w Gdańsku.

Gale, F. 1997. Direct farm marketing as a rural development tool. *Rural Development Perspectives*, 12(2), 19–25

Hugues, F. 2004. Le tourisme durable une organisation du tourisme en milieu rural. *Revue d'Économie Régionale & Urbaine*, 1, 57–80

Kapała, A. 2010. Regulacja prawna agroturystyki uwagi de lege ferenda. *Przegląd Prawa Rolnego*, 7(2), 101–20

Kmita-Dziasek, E. (ed.) 2009. *Marka wiejskiego produktu turystycznego, inicjatywy i inspiracje*, Kraków: Centrum Doradztwa Rolniczego w Brwinowie, Oddz. w Krakowie.

Knecht, D. 2009. *Agroturystyka w agrobiznesie*, Warszawa: Wyd. C.H. Beck.

Kubal, M. and Mika, M. 2012. Agritourism in Poland: the legal model and realities of the market. *Current Issues of Tourism Research*, 2(1), 4–11.

Kurtyka, I. 2008. Turystyka wiejska i agroturystyka jako element zrównoważonego rozwoju powiatu kłodzkiego. *Problemy Zagospodarowania Ziem Górskich*, 55, 175–86.

Mamdy, J.F., Guillot, M. and Disez, N. 2007. Quel avenir pour l'hébergement touristique rural? *Téoros*, 26(3), 56–66.

Marcotte, P., Bourdeau, L. and Doyon, M. 2006. Agrotourisme, agritourisme et tourisme à la ferme ? Une analyse comparative. *Téoros*. 25(3), 59–67.

Mazurski, K.R. 2010. Miejsce i rola agroturystyki we współczesnej turystyce, in *Uwarunkowania integracji w agroturystyce*, edited by D. Knecht, Wrocław: WSZ Edukacja, 10–14.

Palicha, P. (ed.) 2009. *Marka wiejskiego produktu turystycznego*, Gdynia: Wydawnictwo Akademii Morskiej w Gdyni.

Pałka, E. 2008. Przestrzeń agroturystyczna w wielofunkcyjnym regionie gór świętokrzyskich, *Problemy Zagospodarowania Ziem Górskich*, 55, 145–53.

Phillip, S., Hunter, C. and Blackstock, K. 2010. A typology for defining agritourism. *Tourism Management*, 31(6), 754–58.

Pimbert, M. et al. 2001. Global restructuring, agri-food systems and livelihoods, *IIED Gatekeeper Series*, 100.

Rachwał, T. and Zdon-Korzeniowska, M. 2011. Turystyka w warunkach światowego kryzysu gospodarczego. *Prace Komisji Geografii Przemysłu Polskiego Towarzystwa Geograficznego*, 18, 116–28.

Rachwał, T. 2005a. Kształtowanie postaw uczniów na lekcjach przedsiębiorczości, in *Przedsiębiorczość a współczesne wyzwania cywilizacyjne*, edited by Z. Zioło and T. Rachwał, series: Przedsiębiorczość-Edukacja, 1, Kraków: Wydawnictwo Nowa Era, 137–44.

Rachwał, T. 2005b. *Podstawy przedsiębiorczości. Słownik*. 2nd Edition, Warszawa: Nowa Era.

Rachwał, T. 2012. *Projekt „Jestem przedsiębiorczy" a założenia nowej podstawy programowej kształcenia w ramach przedmiotów „podstawy przedsiębiorczości" i „ekonomia" w praktyce, Seminarium dla dyrektorów*. Available at: http://www.jestemprzedsiebiorczy.pl/att/Rachwal_artykul_Seminarium_dla_dyrektorow_6-06-12_tekst.pdf [accessed: December 12, 2012].

Sikora, J., 2012. *Agroturystyka. Przedsiębiorczość na obszarach wiejskich.* Warszawa: Wydawnictwo C.H. Beck.

Sznajder, M. and Przezbórska, L. 2006. *Agroturystyka*, Warszawa: PWE.

Turystyka wiejska, w tym agroturystyka, jako element zrównoważonego i wielofunkcyjnego rozwoju obszarów wiejskich. 2012. Warszawa: Agrotec Polska and Instytut Geografii i Przestrzennego Zagospodarowania PAN.

Ustawa z dnia 2 lipca 2004 r. *o swobodzie działalności gospodarczej* (with amendments).

Ustawa z dnia 26 lipca 1991 r. *o podatku dochodowym od osób fizycznych* (with amendments).

van der Ploeg, J. D., Renting, H., Brunori, G., Knickel, K., Mannion, J., Marsden, T., De Roest, K., Sevilla-Guzmán, E. and Ventura, F., 2000, Rural development: from practices and policies towards theory. *Sociologia Ruralis*, 40, 391–408.

Wiatrak, A. 2003. Baza agroturystyczna w Polsce i uwarunkowania jej rozwoju. *Zeszyty Naukowe Akademii Rolniczej w Krakowie*, 402(90), 9–18.

Wojciechowska, J. 2009. *Procesy i uwarunkowania rozwoju agroturystyki w Polsce*. Łódź: Wyd. Uniwersytetu Łódzkiego.

Wojciechowska, J, 2010. Agroturystyka—signum polskiej turystyki. *Acta Scientiarum Polonorum, Oeconomia* 9(4), 597–606

Zioło, Z. 2007. Rola przedsiębiorczości w aktywizacji gospodarczej—zarys modelu, in *Rola przedsiębiorczości w aktywizacji gospodarczej*, edited by Z. Zioło and T. Rachwał, series: Przedsiębiorczość-Edukacja, 3, Kraków-Warszawa: Wydawnictwo Nowa Era, 10–17.

Index

For Product Safety Concerns and Information please contact our EU representative GPSR@taylorandfrancis.com Taylor & Francis Verlag GmbH, Kaufingerstraße 24, 80331 München, Germany

Batch number: 10725973

Printed by Printforce, the Netherlands